LONDON MATHEMATICAL SOCIETY STUDENT TEXTS

Managing editor: Professor W. Bruce, Department of Mathematics
University of Liverpool, United Kingdom

7 *The theory of evolution and dynamical systems*, J. HOFBAUER & K. SIGMUND
8 *Summing and nuclear norms in Banach space theory*, G. J. O. JAMESON
9 *Automorphisms of surfaces after Nielson and Thurston*, A. CASSON & S. BLEILER
11 *Spacetime and singularities*, G. NABER
12 *Undergraduate algebraic geometry*, M. REID
13 *An introduction to Hankel operators*, J. R. PARTINGTON
15 *Presentations of groups, second edition*, D. L. JOHNSON
17 *Aspects of quantum field theory in curved spacetime*, S. A. FULLING
18 *Braids and coverings: Selected topics*, V. LUNDSGAARD HANSEN
19 *Steps in commutative algebra*, R. Y. SHARP
20 *Communication theory*, C. M. GOLDIE & R. G. E. PINCH
21 *Representations of finite groups of Lie type*, F. DIGNE & J. MICHEL
22 *Designs, graphs, codes, and their links*, P. J. CAMERON & J. H. VAN LINT
23 *Complex algebraic curves*, F. KIRWAN
24 *Lectures on elliptic curves*, J. W. S. CASSELS
25 *Hyperbolic geometry*, B. IVERSEN
26 *An introduction to the theory of L-functions and Eisenstein series*, H. HIDA
27 *Hilbert space: compact operators and the trace theorem*, J. R. RETHERFORD
28 *Potential theory in the complex lane*, T. RANSFORD
29 *Undergraduate commutative algebra*, M. REID
31 *The Laplacian on a Riemannian manifold*, S. ROSENBERG
32 *Lectures on Lie groups and Lie algebras*, R. CARTER, G. SEGAL & I. MACDONALD
33 *A primer of algebraic* D-*modules*, S. C. COUTINHO
34 *Complex algebraic surfaces*, A. BEAUVILLE
35 *Young tableaux*, W. FULTON
37 *A mathematical introduction to wavelets*, P. WOJTASZCZYK
38 *Harmonic maps, loop groups and integrable systems*, M. GUEST
39 *Set theory for the working mathematician*, K. CIESIELSKI
40 *Ergodic theory and dynamical systems*, M. POLLICOTT & M. YURI
41 *The algorithmic resolution of diophantine equations*, N. P. SMART
42 *Equilibrium states in ergodic theory*, G. KELLER
43 *Fourier analysis on finite groups and applications*, A. TERRAS
44 *Classical invariant theory*, P. J. OLVER
45 *Permutation groups*, P. J. CAMERON
46 *Riemann surfaces: A primer*, A. BEARDON
47 *Introductory lectures on rings and modules*, J. BEACHY
48 *Set theory*, A. HÁJNAL & P. HAMBURGER
49 *An introduction to* K-*theory for* C*-*algebras*, M. RØRDAM, F. LARSEN & N. LAUSTSEN
50 *A brief guide to algebraic number theory*, H. P. F. SWINNERTON-DYER
51 *Steps in commutative algebra*, R. Y. SHARP
52 *Finite Markov chains and algorithmic applications*, O. HÄGGSTRÖM
53 *The prime number theorem*, G. J. O. JAMESON
54 *Topics in graph automorphisms and reconstruction*, J. LAURI & R. SCAPELLATO
55 *Elementary number theory, group theory and Ramanujan graphs*, G. DAVIDOFF, P. SARNAK & A. VALETTE
56 *Logic, induction and sets*, T. FORSTER
57 *Introduction to Banach algebras, operators, and harmonic analysis*, G. DALES, P. AIENA, J. ESCHMEIER, K. LAURSEN & G. WILLIS
58 *Computational algebraic geometry*, H. SCHENCK

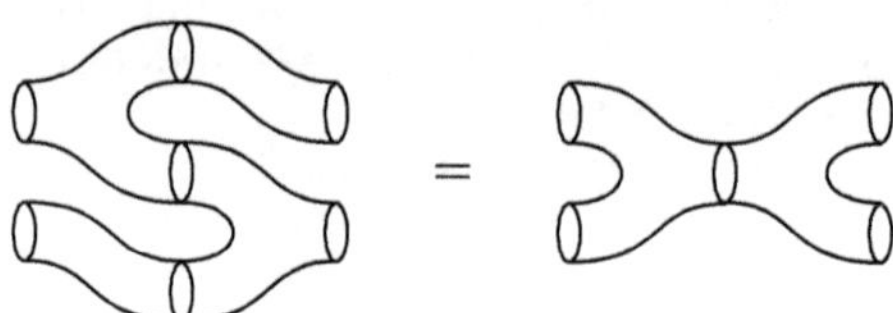

FRONTISPIECE *The topological expression of the main axiom of a Frobenius algebra.* This figure illustrates the content of the main theorem, and captures the whole spirit of the book.

Frobenius Algebras and 2D Topological Quantum Field Theories

JOACHIM KOCK

CAMBRIDGE UNIVERSITY PRESS
Cambridge, New York, Melbourne, Madrid, Cape Town,
Singapore, São Paulo, Delhi, Tokyo, Mexico City

Cambridge University Press
The Edinburgh Building, Cambridge CB2 8RU, UK

Published in the United States of America by
Cambridge University Press, New York

www.cambridge.org
Information on this title: www.cambridge.org/9780521540315

First published 2004

A catalogue record for this publication is available from the British Library

Library of Congress Cataloguing in Publication data
Kock, Joachim, 1967–
Frobenius algebras and 2D topological quantum field theories / Joachim Kock.
p. cm. – (London Mathematical Society student texts; 59)
Includes bibliographical references and index.
ISBN 0 521 83267 5 – ISBN 0 521 54031 3 (paperback)
1. Frobenius algebras. 2. Topological fields. 3. Quantum field theory. I. Title: Frobenius algebras and 2D topological quantum field theories. II. Title. III. Series.
QA251.5.K63 2003
512′.24–dc21 2003055190

ISBN 978-0-521-83267-0 Hardback
ISBN 978-0-521-54031-5 Paperback

For fun

Preface

The text centres around notions of Frobenius structure which in recent years have drawn some attention in topology, physics, algebra, and computer science. In topology the structure arises in the category of 2-dimensional oriented cobordisms (and their linear representations, which are 2-dimensional topological quantum field theories) – this is the subject of the first chapter. The main result here (due to Abrams [1]) is a description in terms of generators and relations of the monoidal category ***2Cob***. In algebra, the structure manifests itself simply as Frobenius algebras, which are treated carefully in Chapter 2. The main result here is a characterisation of Frobenius algebras in terms of comultiplication which goes back to Lawvere [32] and was rediscovered by Quinn [43] and Abrams [1]. The main result of these notes is that these two categories are equivalent: the category of 2-dimensional topological quantum field theories and the category of commutative Frobenius algebras. This result is due to Dijkgraaf [16], further details of the proof having been provided by Quinn [43], Dubrovin [19], and Abrams [1]. The notions from category theory needed in order to express this rigorously (monoidal categories and their linear representations) are developed from an elementary level in Chapter 3. The categorical viewpoint allows us to extract the essence of what is going on in the first two chapters, and prove a natural generalisation of the theorem. To arrive at this insight, we carefully review the classical fact that the simplex category Δ is the free monoidal category on a monoid. (This means in particular that there is an equivalence of categories between the category of algebras and the category of 'linear representations' of Δ.) Now the notion of a Frobenius object in a monoidal category is introduced, and the promised generalisation of the theorem (main result of Chapter 3) states that ***2Cob*** is the free symmetric monoidal category on a commutative Frobenius object.

For more details on the mathematical content, see the Introduction.

The target. The book is based on notes prepared for an intensive two-week mini-course for advanced undergraduate students, given in the UFPE Summer School, Recife, Brazil, in January 2002. The prerequisites are modest: the students of the mini-course were expected to have followed these three standard courses taught at Brazilian universities: one on *differential topology*, one on *algebraic structures* (groups and rings) and one *second course in linear algebra*. From topology we need just some familiarity with the basic notions of differentiable manifolds; from algebra we need basic notions of rings and ideals, groups and algebras; and first and foremost the reader is expected to be familiar with tensor products and hom sets. Usually the course *algebraic structures* contains an introduction to categories and functors, but not enough to get acquainted with the categorical way of thinking and appreciate it; the exposition in this text is meant to take this into account. The basic definitions are given in an appendix, and the more specialised notions are introduced with patience and details, and with many examples – and hopefully the interplay between topology and algebra will provide the appreciation of the categorical viewpoint.

In a wider context these notes are targeted at undergraduate students with a similar background, as well as graduate students of all areas of mathematics. Experienced mathematicians and experts in the field will sometimes be bored by the amount of detail presented, but it is my hope the drawings will keep them awake.

The aim. At an immediate level, the aim of these notes is simply to expose some delightful and not very well known mathematics where a lot of figures can be drawn: a quite elementary and very nice interaction between topology and algebra – and rather different in flavour from what one learns in a course in algebraic topology. On a deeper level, the aim is to convey an impression of unity in mathematics, an aspect which is often hidden from students until later in their mathematical apprenticeship. Finally, perhaps the most important aim is to use this as motivation for category theory, and specifically to serve as an introduction to monoidal categories.

Admittedly, the main theorem is not a particularly useful tool that the students will draw upon again and again throughout their mathematical career, and one could argue that the time would be better spent on a course on group representations or distributions, for instance. But after all, this is a summer school (and this is Brazil!): maximising the throughput is not our main concern – the wonderful relaxed atmosphere I know from previous summer schools in Recife is much more important – I hope the students when they go to the beach in the weekend will make drawings of 2-dimensional cobordisms in the sand! (I think they would not take orthogonality relations or Fourier transforms with them to the beach. . .)

What the lectures are meant to give the students are rather some techniques and viewpoints, and in the end this categorical perspective reduces the main theorem to a special case of general principles. A lot of emphasis is placed on universal properties, symmetry, distinction between structure and property, distinction between identity and natural isomorphism, the interplay between graphical and algebraic approaches to mathematics – as well as reflection on the nature of the most basic operations of mathematics: multiplication and addition. Getting acquainted with such categorical viewpoints in mathematics is certainly a good investment.

Finally, to be more concrete, the techniques learned in this course should constitute a good primer for going into quantum groups or knot theory.

The source – acknowledgements. The idea of these notes originated in a workshop I led at KTH, Stockholm, in 2000, whose first part was devoted to understanding the paper of Abrams [1] (corresponding more or less to Chapters 1 and 2 of this text). I am thankful for the contributions of the core participants of the workshop: ***Carel*** Faber, ***Helge*** Måkestad, ***Mats*** Boij, and ***Michael*** Shapiro, and in particular to ***Dan*** Laksov, for many fruitful discussions about Frobenius algebras.

The more categorical viewpoint of Chapter 3 was influenced by the people I work with here in Nice; I am indebted in particular to ***André*** Hirschowitz and ***Bertrand*** Toën. I have also benefited from discussions and email correspondence with ***Arnfinn*** Laudal, ***Göran*** Fors, ***Jan*** Gorski, ***Jean-Louis*** Cathélineau, ***John*** Baez, and ***Pedro*** Ontaneda, all of whom are thanked. I am particularly indebted to ***Anders*** Kock, ***Peter*** Johnson, and ***Tom*** Leinster for many discussions and helpful emails, and for carefully reading preliminary versions of the manuscript, pointing out grim errors, annoying inaccuracies, and misprints.

Israel Vainsencher, ***Joaquim*** Roé, ***Ramón*** Mendoza, and ***Sérgio*** Santa Cruz also picked up some misprints – thanks. My big sorrow about these notes is that I do not understand the physics behind it all, in spite of a great effort by ***José*** Mourão to explain it to me – I am grateful to him for his patience.

During the redaction of these notes I have reminisced about maths classes in primary school, and some of the figures are copied from my very first maths books. Let me take the opportunity to thank ***Marion*** Kuhlmann and ***Jørgen*** Skaftved for the mathematics they taught me when I was a child.

During my work with this subject and specifically with these notes, I have been supported by ***The National Science Research Council of Denmark***,

The Nordic Science Research Training Academy ***NorFA***, and (currently) a Marie Curie Fellowship from ***The European Commission***. In neither case was I supposed to spend so much time with Frobenius algebras and topological quantum field theories – it is my hope that these notes, as a concrete outcome of the time spent, do it justice to some extent.

I am indebted to my wife ***Andrea*** for her patience and support.

Last but not least, I wish to thank the organisers of the Summer School in Recife – in particular ***Letterio*** Gatto – for inviting me to give this mini-course, which in addition to being a very dear opportunity to come back to Recife – *Voltei, Recife! foi a saudade que me trouxe pelo braço* – has also been a welcome incentive to work out the details of this material and learn a lot of mathematics.

Feedback is most welcome. Please point out mathematical errors or misunderstandings, misleading viewpoints, unnecessary pedantry, or things that should be better explained; typos, mispellings, bad English, TEX-related issues. I intend to keep a list of errata on my web site.

The original LATEX source files were prepared in alpha. The figures were coded with the texdraw package, written by Peter Kabal. The diagrams were set using the diagrams package of Paul Taylor, except for the curved arrows which were coded by hand.

Contents

General conventions

We consistently write composition of functions (or arrows) from the left to the right: given functions (or arrows)

$$X \xrightarrow{f} Y \xrightarrow{g} Z$$

we denote the composite fg. Similarly, we put the symbol of a function to the right of its argument, writing for example

$$\begin{aligned} f : X &\longrightarrow Y \\ x &\longmapsto xf. \end{aligned}$$

Introduction

In this introduction we briefly explain the words of the title of these notes, give a sketch of what we are going to do with these notions, and outline the viewpoint we will take in order to understand the structures. In the course of this introduction a lot of other words will be used which are probably no more familiar than those they are meant to explain – but don't worry: in the main text, all these words are properly defined and carefully explained...

Frobenius algebras. A Frobenius algebra is a finite-dimensional algebra equipped with a nondegenerate bilinear form compatible with multiplication. (Chapter 2 is all about Frobenius algebras.) Examples are matrix rings, group rings, the ring of characters of a representation, and Artinian Gorenstein rings (which in turn include cohomology rings, local rings of isolated hypersurface singularities...)

In algebra and representation theory such algebras have been studied for a century, along with various related notions – see Curtis and Reiner [15].

Frobenius structures. During the past decade, Frobenius algebras have shown up in a variety of topological contexts, in theoretical physics and in computer science. In physics, the main scenery for Frobenius algebras is that of topological quantum field theory (TQFT), which in its axiomatisation amounts to a precise mathematical theory. In computer science, Frobenius algebras arise in the study of flowcharts, proof nets, circuit diagrams...

In any case, the reason Frobenius algebras show up is that this is essentially a topological structure: it turns out that the axioms for a Frobenius algebra can be given completely in terms of graphs – or as we shall do, in terms of topological surfaces.

Frobenius algebras are just algebraic representations of this structure – the goal of these notes is to make all this precise. We will focus on topological quantum field theories – and in particular on dimension 2. This is by far the best picture of Frobenius structures since the topology is explicit, and since there is no additional structure to complicate things. In fact, the main theorem of these notes states that there is an equivalence of categories between that of 2-dimensional TQFTs and that of commutative Frobenius algebras.

(There will be no further mention of computer science in these notes.)

Topological quantum field theories. In the axiomatic formulation (due to Atiyah [5]), an n-dimensional topological quantum field theory is a rule $\mathscr{A}$ which to each closed oriented manifold Σ (of dimension $n-1$) associates a vector space $\Sigma\mathscr{A}$, and to each oriented n-manifold whose boundary is Σ associates a vector in $\Sigma\mathscr{A}$. This rule is subject to a collection of axioms which express that topologically equivalent manifolds have isomorphic associated vector spaces, and that disjoint unions of manifolds go to tensor products of vector spaces, etc.

Cobordisms. The clearest formulation is in categorical terms. First one defines a category of cobordisms $\boldsymbol{nCob}$: the objects are closed oriented $(n-1)$-manifolds, and an arrow from Σ to Σ' is an oriented n-manifold M whose 'in-boundary' is Σ and whose 'out-boundary' is Σ'. (The cobordism M is defined up to diffeomorphism rel the boundary.) The simplest example of a cobordism is the cylinder $\Sigma \times I$ over a closed manifold Σ, say a circle. It is a cobordism from one copy of Σ to another.

Here is a drawing of a cobordism from the union of two circles to one circle:

Composition of cobordisms is defined by gluing together the underlying manifolds along common boundary components; the cylinder $\Sigma \times I$ is the identity arrow on Σ. The operation of taking disjoint union of manifolds and cobordisms gives this category *monoidal structure* – more about monoidal categories later. On the other hand, the category $\boldsymbol{Vect}_{\Bbbk}$ of vector spaces is monoidal under tensor products.

Now the axioms amount to saying that a TQFT is a (symmetric) monoidal functor from $\boldsymbol{nCob}$ to $\boldsymbol{Vect}_{\Bbbk}$. This is also called a linear representation of $\boldsymbol{nCob}$.

So what does this have to do with Frobenius algebras? Before we come to the relation between Frobenius algebras and 2-dimensional TQFTs, let us make a couple of remarks on the motivation for TQFTs.

Physical interest in TQFTs comes mainly from the observation that TQFTs possess certain features one expects from a theory of quantum gravity. It serves as a baby model in which one can do calculations and gain experience before embarking on the quest for the full-fledged theory, which is expected to be much more complicated. Roughly, the closed manifolds represent *space*, while the cobordisms represent *space-time*. The associated vector spaces are then the *state spaces*, and an operator associated to a space-time is the time-evolution operator (also called transition amplitude, or Feynman path integral). That the theory is topological means that the transition amplitudes do not depend on any additional structure on space-time (like Riemannian metric or curvature), but only on the topology. In particular there is no time-evolution along cylindrical space-time. That disjoint union goes to tensor product expresses the common principle in quantum mechanics that the state space of two independent systems is the tensor product of the two state spaces.

(No further explanation of the relation to physics will be given – the author of these notes recognises he knows nearly nothing of this aspect. The reader is referred to Dijkgraaf [17] or Barrett [11], for example.)

Mathematical interest in TQFTs stems from the observation that they produce invariants of closed manifolds: an n-manifold without boundary is a cobordism from the empty $(n-1)$-manifold to itself, and its image under $\mathscr{A}$ is therefore a linear map $\Bbbk \to \Bbbk$, i.e. a scalar. It was shown by Witten how TQFT in dimension 3 is related to invariants of knots and the Jones polynomial – see Atiyah [6].

The viewpoint of these notes is different however: instead of developing TQFTs in order to describe and classify manifolds, we work in dimension 2 where a complete classification of surfaces already exists; we then use this classification to describe TQFTs!

Cobordisms in dimension 2. In dimension 2, 'everything is known': since surfaces are completely classified, one can also describe the cobordism category completely. Every cobordism is obtained by composing the following basic building blocks (each with the in-boundary drawn to the left):

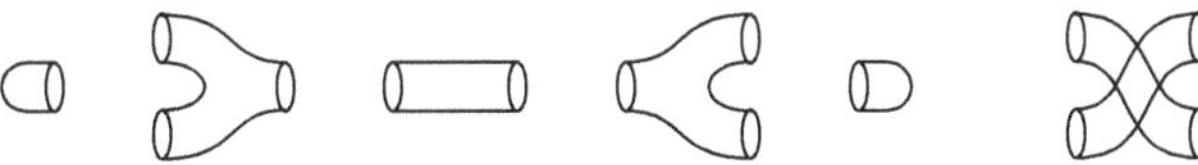

Two connected cobordisms are equivalent if they have the same genus and the same number of in- and out-boundaries. This gives a bunch of relations, and a complete description of the monoidal category ***2Cob*** in terms of generators and relations. Here are two examples of relations that hold in ***2Cob***:

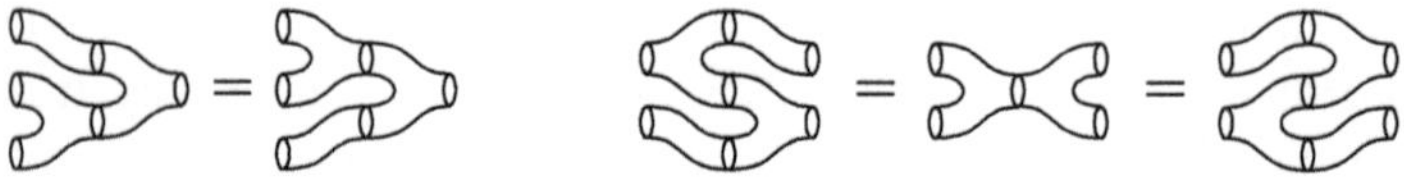

(1)

These equations express that certain surfaces are topologically equivalent rel the boundary.

Topology of some basic algebraic operations. Some very basic principles are in play here: 'creation', 'coming together', 'splitting up', 'annihilation'. These principles have explicit mathematical manifestations as algebraic operations:

Principle	Feynman diagram	2D cobordism	Algebraic operation (in a $\Bbbk$-algebra A)	
merging	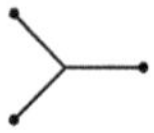		multiplication	$A \otimes A \to A$
creation			unit	$\Bbbk \to A$
splitting	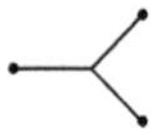		comultiplication	$A \to A \otimes A$
annihilation			counit	$A \to \Bbbk$

Note that in the intuitive description there is a notion of time involved which accounts for the distinction between coming-together and splitting-up – or perhaps 'time' is too fancy a word, but at least there is a notion of *start* and *finish*. Correspondingly, in the algebraic or categorical description the notion of morphism involves a direction: morphisms are *arrows*, and they have well defined *source* and *target*.

It is an important observation from category theory that many algebraic structures admit descriptions purely in terms of arrows (instead of referring to elements) and commutative diagrams (instead of equations among elements). In particular, this is true for the notion of an *algebra*: an algebra is a vector space A equipped with two maps $A \otimes A \to A$ and $\Bbbk \to A$, satisfying the associativity axiom and the unit axiom. Now according to the above dictionary, the left-hand relation of Equation (1) is just the topological expression of associativity! Put in other words, the associativity equation has topological content: it expresses the topological equivalence of two surfaces (or two graphs).

It gives sense to other operations, like merging (or splitting) three particles: it makes no difference whether we first merge two of them and then merge the result with the third, or whether we merge the last two with the first. From the viewpoint of graphs, the basic axiom (equivalent to Equation (1)) is that two vertices can move past each other:

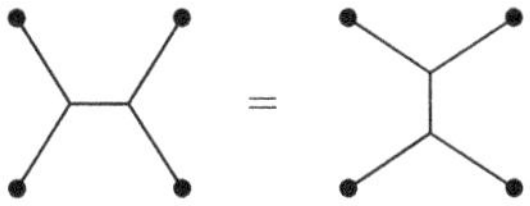

Frobenius algebras. In order to relate this to Frobenius algebras the definition given in the beginning of this introduction is not the most convenient. It turns out one can characterise a Frobenius algebra as follows: it is an algebra (multiplication denoted ) which is simultaneously a *coalgebra* (comultiplication denoted ) with a certain compatibility condition between and  . This compatibility condition is exactly the right-hand relation drawn in Equation (1). (Note that by the dictionary, this is just a graphical expression of a precise algebraic requirement.) In fact, the relations that hold in ***2Cob*** correspond precisely to the axioms of a commutative Frobenius algebra. This comparison leads to the main theorem:

Theorem. *There is an equivalence of categories*

$$\textbf{\textit{2TQFT}} \simeq \textbf{cFA},$$

given by sending a TQFT to its value on the circle (the unique closed connected 1*-manifold).*

So in this sense, we can say, if we want, that Frobenius algebras are the same thing as linear representations of ***2Cob***.

The idea of the proof is this: let A be the image of the circle, under a TQFT $\mathscr{A}$. Now $\mathscr{A}$ sends each of the generators of ***2Cob*** to a linear map between tensor powers of A, just as tabulated above. The relations which hold in ***2Cob*** are preserved by $\mathscr{A}$ (since $\mathscr{A}$ by definition is a monoidal functor) and in its target category ***Vect*** they translate into the axioms for a commutative Frobenius algebra! (Conversely, every commutative Frobenius algebra can be used to define a 2-dimensional TQFT.)

Monoidal categories. As mentioned, just in order to define the category ***TQFT*** we need the notion of monoidal categories. In fact, monoidal categories is the best framework to understand all the concepts described above. The notion of associative multiplication with unit is precisely what the abstract concept of *monoid* encodes – and monoids live in monoidal categories.

The prime example of a monoidal category is the category $\textbf{\textit{Vect}}_{\Bbbk}$ of vector spaces and tensor products, with the ground field as neutral object. In general a monoidal category is a category equipped with some sort of 'product' like $\otimes$ or $\coprod$, satisfying certain properties. This 'product' serves as background for defining the multiplication maps, i.e. defining monoids: a monoid in $(\textbf{\textit{Vect}}_{\Bbbk}, \otimes, \Bbbk)$ is precisely a $\Bbbk$-algebra A, since the multiplication map is described as a $\Bbbk$-linear map $A \otimes A \to A$, etc. Another example of a monoid is the circle in ***2Cob***...

The simplex category Δ and what it means to monoids and algebras. There is a little monoidal category which bears some similarity with ***2Cob***: the simplex category Δ is roughly the category of finite ordered sets and order-preserving maps. It is a monoidal category under disjoint union. To be more precise, the objects of Δ are $\mathbf{n} = \{0, 1, 2, \ldots, n-1\}$, one for each $n \in \mathbb{N}$, and the arrows are the maps $f : \mathbf{m} \to \mathbf{n}$ such that $i \leq j \Rightarrow if \leq jf$. There are several other descriptions of this important category – one is in graphical terms, and reveals it as a subcategory of ***2Cob***. The object $\mathbf{1}$ is a monoid in Δ, and in a sense Δ is the smallest possible monoidal category which contains a nontrivial monoid. In fact the following universal property is shown to hold: *every monoid in any monoidal category* **V** *is the image of* **1** *under a unique monoidal functor* $\Delta \to \mathbf{V}$. This is to say that Δ is the free monoidal category containing a monoid. In particular, $\Bbbk$-algebras can be interpreted as 'linear representations' of Δ.

Observing that Δ can be described graphically, we see that this result is of exactly the same type as our main Theorem.

Frobenius objects. Once we have taken the step of abstraction from $\Bbbk$-algebras to monoids in an arbitrary monoidal category, it is straightforward to define the notion of Frobenius object in a monoidal category: it is an object equipped with four maps as those listed in the table, and with the compatibility condition expressed in Equation (1). In certain monoidal categories, called symmetric, it makes sense to ask whether a monoid or a Frobenius object is commutative, and of course these notions are defined in such a way that commutative Frobenius objects in $\textbf{\textit{Vect}}_{\Bbbk}$ are precisely commutative Frobenius algebras.

Universal Frobenius structure. With these general notions, generalisation of the Theorem is immediate: all the arguments of the proof do in fact carry over to the setting of an arbitrary (symmetric) monoidal category, and we find that ***2Cob*** is the free symmetric monoidal category containing a commutative Frobenius object. This means that *every commutative Frobenius object in any symmetric monoidal category* **V** *is the image of the circle under a unique symmetric monoidal functor from* ***2Cob***.

Since the proof of this result is the same as the proof of the original theorem, this is the natural generality of the statement. The interest in this generality is that it actually includes many natural examples of TQFTs which could not fit into the original definition. For example, in our treatment of Frobenius algebras in Chapter 2 we will see that cohomology rings are Frobenius algebras in a natural way, but typically they are not commutative but only graded-commutative. For this reason they cannot support a TQFT in the strict sense. But if instead of the usual symmetric monoidal category ***Vect*** we take for example the category of graded vector spaces with 'super-symmetry' structure, then all cohomology rings can support a TQFT (of this slightly generalised sort).

It is the good generalised version of the main theorem that makes this clear. In many sources on TQFTs, the questions of symmetry are swept under the carpet and the point about 'super-symmetric' TQFTs is missed.

In these notes, the whole question of symmetry is given a rather privileged rôle. The difficult thing about symmetry is to avoid mistaking it for identity! For example, for the cartesian product $\times$ (which is an important example of a monoidal structure), it is *not* true that $X \times Y = Y \times X$. What is true is that there is a natural isomorphism between the two sets (or spaces). Similar observations are due for disjoint union $\coprod$, and tensor product $\otimes$. . . While it requires some pedantry to treat symmetry properly, it is necessary in order to understand the super-symmetric examples just mentioned.

Organisation of these notes. The notes are divided into three chapters each of which should be read before the others! The first chapter is about topology – cobordisms and TQFTs; Chapter 2 is about algebra – Frobenius algebras; and Chapter 3 is mostly category theory. The reader is referred to the Contents for more details on where to find what.

Although the logical order of the material is not completely linear, hopefully the order is justified pedagogically: we start with geometry! – the concrete and palpable – and then we gradually proceed to more abstract subjects (or should we say: more abstract aspects of our subject), helped by drawings and intuition provided by the geometry. With the experience gained with these investigations we get ready to try to understand the abstract structures behind. The ending is about very abstract concepts and objects with universal properties, but we can cope with that because we know the underlying geometry – in fact we show that this very abstract thing with that universal property is precisely the cobordism category we described so carefully in Chapter 1.

Exercises. Each section ends with a collection of exercises of varying level and interest. Most of them are really easy, and the reader is encouraged to do them all. A few of them are considered less straightforward and have been marked with a star.

Further reading. My great sorrow about these notes is that I do not understand the physical background or interpretation of TQFTs. The physically inclined reader must resort to the existing literature, for example Atiyah's book [6] or the notes of Dijkgraaf [17]. I would also like to recommend John Baez's web site [8], where a lot of references can be found.

Within the categorical viewpoint, an important approach to Frobenius structures which has not been touched upon is the 2-categorical viewpoint, in terms of monads and adjunctions. This has recently been exploited to great depth by Müger [39]. Again, a pleasant introductory account is given by Baez [8], TWF 174 (and 173).

Last but not least, I warmly recommend the lecture notes of Quinn [43], which are detailed and go in depth with concrete topological quantum field theories.

1

Cobordisms and topological quantum field theories

Summary

In the first section we recall some basic notions of manifolds with boundary and orientations, and Morse functions. We introduce the slightly nonstandard notion of *in-boundary* and *out-boundary*, which is particularly convenient for the treatment of cobordisms.

Section 1.2 is devoted to the basic theory of oriented cobordisms. Roughly a cobordism between two closed $(n-1)$-manifolds is an n-manifold whose boundary is made up of the two $(n-1)$-manifolds. We describe what it means for two cobordisms to be equivalent. Next we introduce the *decomposition* of a cobordism, which amounts to cutting up along a closed codimension-1 submanifold, obtaining two cobordisms. Finally we state the axioms for a topological quantum field theory (TQFT) in the style of Atiyah [5]: it is a way of associating vector spaces and linear maps to $(n-1)$-manifolds and cobordisms, respecting decompositions and disjoint union. A special decomposition of the cylinder shows that a vector space which is image of a TQFT comes equipped with a nondegenerate bilinear pairing, in a strong sense, which in particular forces the vector space to be finite dimensional.

In Section 1.3 we assemble the manifolds and cobordisms into a category ***nCob***. In order to have a well defined composition we must pass to a quotient, identifying equivalent cobordisms. The identity arrows are the cylinder classes. Then we start discussing the monoidal structure: disjoint union of cobordisms. With this terminology we can define a TQFT as a (symmetric) monoidal functor from ***nCob*** to $\textbf{\textit{Vect}}_{\Bbbk}$. (The definition and basic properties of monoidal categories are given in Chapter 3.)

Finally in Section 1.4 we specialise to dimension 2. Here we can give a complete description of ***2Cob*** in terms of generators and relations for a monoidal category. These results depend on the classification theorem for topological surfaces.

The notion of cobordism goes back to Pontryagin and Thom [47] in the 1950s. Topological quantum field theories were introduced by Witten [50], and the mathematical axiomatisation was soon after proposed by Atiyah [5] (1989). The description of ***2Cob*** in terms of generators and relations was first given explicitly by Abrams [1] (1995), but the proof is already sketched in Quinn [43], and most likely it goes further back.

1.1 Geometric preliminaries

The reader is expected to be familiar with the most basic notions of differentiable manifolds, their tangent bundles, smooth maps, and their differentials. Our main reference will be Hirsch [27]. A more elementary introduction which emphasises the concepts used here is Wallace [48]. In this section we just collect some crucial notions, establish terminology and notation, and give some basic examples which we will need later on.

Our manifolds are not assumed to be embedded in Euclidean space; an n-manifold is merely a topological space M covered by open sets homeomorphic to $\mathbb{R}^n$. These maps are called coordinate charts, and the collection of all of them is called an atlas. *Smooth* means differentiable of class C^∞; that is, on overlaps between two such charts, the coordinate change functions are differentiable maps of class C^∞ (between subsets of $\mathbb{R}^n$). A smooth structure on M is a maximal smooth atlas. Throughout, *manifold* will mean smooth manifold, i.e. a manifold equipped with a smooth structure. All our manifolds will be compact, but we do not assume them to be connected.

Let us note that we regard the empty set as an n-manifold! – we denote it $\emptyset_n$. This is justified by the observation that every point of $\emptyset_n$ has a neighbourhood homeomorphic to $\mathbb{R}^n$.

Manifolds with boundary

In a usual n-manifold M, every point x has a neighbourhood homeomorphic to $\mathbb{R}^n$. This means that from x you can move a little bit in any direction. This is possible either because M 'curves back and closes up itself' like a circle, a sphere, or a torus, or because M is open, like for example the open disc $\{x \in \mathbb{R}^2 \mid |x| < 1\}$: here the points are not allowed to sit on the boundary, so no matter how close you are to it you can always come a little bit closer.

In a manifold with boundary the 'boundary points' are included, like for example the points on the circumference in $\{x \in \mathbb{R}^2 \mid |x| \leq 1\}$. For such a point, there are directions in which it is impossible to move: there is no neighbourhood homeomorphic to $\mathbb{R}^n$, so we need a new sort of chart. It is practical to require the boundary itself to be a manifold (of dimension $n-1$); this is achieved by allowing half-space charts in the sense of the next paragraph.

Throughout this book there are hundreds of pictures of manifolds with boundary. Most of them look like tubes – the boundary consists of the ends of the tubes.

1.1.1 Half-spaces. A *half-space* is a set of the form

$$H^n = \{x \in \mathbb{R}^n \mid x\Lambda \geq 0\}$$

where $\Lambda : \mathbb{R}^n \to \mathbb{R}$ is a nonzero linear map. (Recall our general convention (page xiv): $x\Lambda$ means the value of Λ at x.) (Half-spaces do not exist in dimension 0, since there are no nonzero linear maps on $\mathbb{R}^0$.)

The topology of H^n is induced from that of $\mathbb{R}^n$. The tangent space of a point on ∂H^n is a full vector space $\mathbb{R}^n$ (not something half!), so it makes sense to speak of differentiability of functions on H^n.

The *boundary* of H^n is the nullspace of Λ:

$$\partial H^n := \mathrm{Null}(\Lambda) = \{x \in \mathbb{R}^n \mid x\Lambda = 0\} \simeq \mathbb{R}^{n-1}.$$

Note that a point on ∂H^n has no neighbourhood in H^n homeomorphic to an open set in $\mathbb{R}^n$.

1.1.2 Manifolds with boundary. An *n-manifold with boundary* is a topological space M covered by open sets each of which is homeomorphic to an open set in H^n. A point $x \in M$ is a *boundary point* if in some coordinate chart it corresponds to a point on the boundary ∂H^n. Then it will also correspond to boundary points in every other chart – these points are characterised as having no neighbourhood homeomorphic to an open set in $\mathbb{R}^n$. The set of all boundary points is covered by open sets of $\mathbb{R}^{n-1} \simeq \partial H^n \subset H^n$, and becomes in this way an $(n-1)$-manifold (without boundary), denoted ∂M.

We require smoothness: all coordinate changes must be smooth maps (between open sets in H^n).

We do not exclude the possibility that the boundary is empty, and in this way every manifold can be considered a manifold with boundary (except in dimension 0 where the notion of manifold with boundary is not defined). Note that every open set in $\mathbb{R}^n$ is homeomorphic to an open set in H^n.

By definition, a *closed manifold* is a compact manifold *without* boundary. (Warning: the adjective 'closed' has another meaning for intervals. A closed interval is one which includes its end-points. In particular, a closed interval is not a closed manifold. Hopefully this will not lead to confusion.)

In most situations, when we talk about a smooth map $f : M \to N$ between manifolds with boundary we will require it to take the boundary of M to the boundary of N. However in some cases this is too restrictive: for example, the inclusion of an interval into a bigger one does not have this property.

1.1.3 Notational convention. For convenience, we consistently denote manifolds with boundary by capital Roman letters (typically M), while manifolds without boundary (and typically in dimension one less) are denoted by capital Greek letters, like Σ.

1.1.4 Examples. A point is a closed 0-manifold. More generally, a finite set of points is a closed 0-manifold.

The unit interval $I = [0, 1] \subset \mathbb{R}^1$ is a 1-manifold with boundary; the boundary has two connected components, namely the points 0 and 1. The circle is a 1-manifold with empty boundary.

The disc $\{x \in \mathbb{R}^2 \mid |x| \leq 1\}$ is a manifold with boundary; the boundary is the circumference.

A common way to obtain manifolds with boundary is to start with a manifold (e.g. a sphere), and remove an open subset bounded by a closed codimension-1 submanifold (e.g. cut away an open disc from the sphere). You could also cut the manifold in two pieces along a closed codimension-1 submanifold Σ (e.g. cut the sphere along equator), then the two pieces become manifolds with boundary – provided Σ is included in each piece, precisely to form its boundary.

1.1.5 Example: cylinders. Let Σ be a manifold without boundary and let I be a manifold with boundary. Then the product manifold $\Sigma \times I$ is a manifold with boundary; the boundary is $\Sigma \times \partial I$. If I is a closed interval $[a, b]$ then we call the product $\Sigma \times I$ a *cylinder* over Σ. The boundary consists of two copies of Σ namely $\Sigma_a := \Sigma \times \{a\}$ and $\Sigma_b := \Sigma \times \{b\}$. The projection $\Sigma \times I \to I$ is a smooth map of manifolds with boundary (it sends the boundary of $\Sigma \times I$ to the boundary of I):

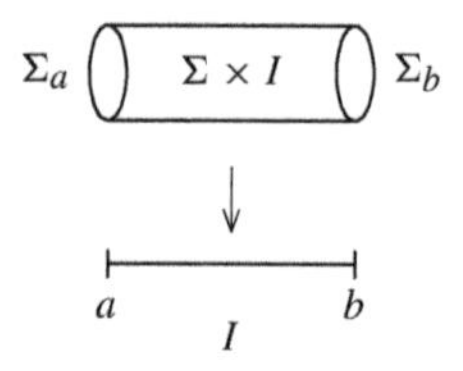

Σ_a $\Sigma \times I$ Σ_b

a b

I

Orientations

We will need to be able to say that some of the boundary components of a manifold are 'in' and some are 'out'. This notion is defined in 1.1.11, and to this end we need the notion of orientation. The technical aspects of orientations presented in the next couple of paragraphs will not really be needed elsewhere in the text.

1.1.6 Orientation of a vector space. Let V be a real vector space of finite dimension. An *orientation* of V is given by associating a sign to each ordered basis, in such a way that two ordered bases have the same sign if and only if the linear transformation taking one to the other has positive determinant. Thus an orientation is specified completely as soon as one ordered basis has been given a sign. So there exist precisely two possible orientations of V.

As an example, the standard orientation of $\mathbb{R}^2$ is the one that declares the standard basis $\left[\binom{1}{0}, \binom{0}{1}\right]$ to be positive – here and on the next couple of pages we use square brackets to denote ordered bases; after that we will not need the notion again.

In the special case where V is the trivial vector space $\{0\}$, there exists only one basis, namely the empty set, so an orientation in this case is given simply by assigning a plus or a minus to this basis – or as we shall put it abusively: an orientation of $V = \{0\}$ is given by a sign attributed to V.

A linear map between oriented vector spaces *preserves* orientation if it takes every positive basis to a positive basis (and thus automatically takes negative bases to negative bases). Note that there are no orientation-preserving maps between $(\{0\}, +)$ and $(\{0\}, -)$.

1.1.7 Oriented manifolds. An *orientation* of a manifold is a smooth choice of orientations of each of its tangent spaces. The smoothness condition is to be understood like this: the differentials of the transition functions (coordinate changes) should all preserve the orientations.

Not every manifold admits an orientation, the most well known examples being the Möbius strip and the real projective plane. If a manifold admits an orientation it is called *orientable*.

If M is orientable and connected there are exactly two possible orientations. If an orientable M is disconnected and has k components, then there are 2^k different orientations. In any case, if M is an oriented manifold, we denote by $\overline{M}$ the same manifold with the opposite orientation.

1.1.8 Special cases. If M is 0-dimensional, an orientation is given by attributing a sign to each point. Indeed, if M is a single point, then the tangent space at this point is the trivial vector space $\{0\}$, which has a unique basis (the empty set): we must give this basis a sign.

The empty manifold $M = \varnothing$ has precisely one orientation. There are no tangent spaces to orient, and for each of them we must associate a sign to each of their bases. So altogether we are counting maps from $\varnothing$ to $\{-, +\}$, and there is just one such map. (This is in agreement with the count 2^k performed above.)

1.1.9 Orientation of a product. If X and Y are oriented manifolds (one of them without boundary), then the product $X \times Y$ acquires an orientation according to the following natural convention: if (x, y) is a point on $X \times Y$ and $[v_1, \ldots, v_n]$ (respectively $[w_1, \ldots, w_m]$) is a positive basis for $T_x X$ (respectively $T_y Y$), then $[v_1, \ldots, v_n, w_1, \ldots, w_m]$ is declared to be a positive basis for $T_{(x,y)}(X \times Y)$.

Hey, if you invert the order of the factors, you get another orientation!?, the reader may object. Well, yes and no: you get 'another' orientation because $X \times Y$ is not the same manifold as $Y \times X$. Of course they are *isomorphic*, and the natural isomorphism is the twist map $X \times Y \stackrel{\sim}{\to} Y \times X,\ (x, y) \mapsto (y, x)$. Now if you compare the two orientations carefully along this isomorphism you will note that they agree!

1.1.10 Example. Let Σ be an closed oriented manifold (with positive basis $[v_1, \ldots, v_m]$ in the tangent space at some point $x \in \Sigma$). Let I denote the unit interval (with standard orientation: $[e_1]$ is a positive basis). Then the product orientation of the cylinder $\Sigma \times I$ has positive basis $[v_1, \ldots, v_m, e_1]$.

1.1.11 In-boundaries and out-boundaries. Let Σ be a closed submanifold of M of codimension 1. Assume both are oriented. At a point $x \in \Sigma$, let $[v_1, \ldots, v_{n-1}]$ be a positive basis for $T_x\Sigma$. A vector $w \in T_x M$ is called a *positive normal* if $[v_1, \ldots, v_{n-1}, w]$ is a positive basis for $T_x M$.

Now suppose Σ is a connected component of the boundary of M; then it makes sense to ask whether the positive normal w points inwards or outwards compared to M – locally the situation is that of a vector in $\mathbb{R}^n$ for which we ask whether it points in or out from the half-space H^n. If a positive normal points inwards we call Σ an *in-boundary*, and if it points outwards we call it an *out-boundary*. To see that this makes sense we have to check that this does not depend on the choice of positive normal (or on the choice of point $x \in \Sigma$). If some positive normal points inwards, it is a fact that every other positive normal at any other point $y \in \Sigma$ points inwards as well. This follows from the fact that the normal bundle $TM\,|_\Sigma\,/T\Sigma$ is a trivial vector bundle on Σ. This in turn is a consequence of the assumption that both M and Σ are orientable (see Hirsch [27], 4.4.2.)

Thus *the* boundary of a manifold M is the union of various in-boundaries and out-boundaries. The in-boundary of M may be empty, and the out-boundary may also be empty. Note that if we reverse the orientation of both M and its boundary Σ, then the notion of what is in-boundary or out-boundary is still the same.

1.1.12 Example. Consider the unit interval with its standard orientation (induced from $\mathbb{R}^1$), and let also the boundary points 0 and 1 be equipped with their standard orientation $+$. Then 0 is an in-boundary and 1 is an out-boundary.

More generally, if Σ is a closed oriented manifold then the cylinder $\Sigma \times I$ (given the product orientation as in 1.1.9) has boundary $\Sigma_0 \coprod \Sigma_1$. Now Σ_0 is an in-boundary and Σ_1 is an out-boundary. (The symbol $\coprod$ denotes the disjoint union; it will be discussed in detail in 1.3.24 and 1.3.25.)

If $\Sigma \subset M$ is a submanifold of codimension 1 which divides M into two parts, then Σ is an out-boundary for one of the parts and an in-boundary for the other.

1.1.13 Example. Consider the unit sphere $S^2 \subset \mathbb{R}^3$, oriented by declaring positive the basis $[E, N]$ (East, North) of the tangent space of S^2 at some point on the Equator Σ (with positive basis $[E]$). Then N is a positive normal of Σ. So Σ is an out-boundary for the southern hemisphere and it is an in-boundary for the northern hemisphere.

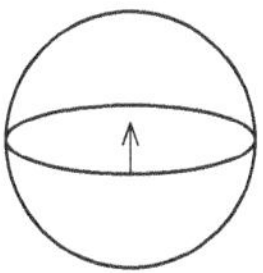

Some vocabulary from Morse theory

The closed interval, say $I = [0, 1]$, is the simplest example of a manifold with boundary. In higher dimension we have seen the cylinders $\Sigma \times I$ as a closely related analogue, and we noted that there is a smooth map $\Sigma \times I \to I$ compatible with the boundaries.

In general, if M is a manifold with boundary it is very useful to have a map like this:

(1.1.14)

A lot of topological properties of M can be detected with such maps. This is what Morse theory is about – a very powerful tool in differentiable topology. In this book we will only need the two or three most basic notions. The next couple of paragraphs just put some technical constraints on those maps, which are necessary for the differentiable machinery to work, but the essential thing to record is just the picture above...

1.1.15 Critical points. Let M be a compact manifold, and consider a smooth map $f : M \to I$ from M to a closed interval $I \subset \mathbb{R}$. A point $x \in M$ is called a *critical point* if the differential df_x is zero. In that case, the image of x under f is called a *critical value* of f. If x is not a critical point we call it a *regular point* for f. If a value is not a critical value we call it a *regular value*; this means that every point in the preimage is a regular point.

In the previous picture, x is a critical point of f, and t is a critical value. All other points of M are regular points (and hence all other points in I are regular values).

1.1.16 Nondegenerate critical points. A critical point x is called nondegenerate if df_x vanishes only to order 1 (i.e. x is a simple zero). More precisely a critical point is *nondegenerate* if in some coordinate system the Hessian matrix

$$\frac{\partial^2 f}{\partial x^i \partial x^j}$$

is nonsingular (this notion does not depend on the coordinate system chosen). Since the Hessian is a real symmetric matrix, all its eigenvalues are real; the *index* of f at x is the number of negative eigenvalues of the Hessian (counted with multiplicity). This number can also be described as the dimension of the largest subspace of $T_x M$ on which the corresponding bilinear form is negative definite.

1.1.17 Examples. If M is a surface and $f : M \to I$ is a smooth function, then a nondegenerate critical point has index 0 if and only if it is a local minimum;

it has index 2 if and only if it is a local maximum; and it has index 1 if and only if it is a saddle point.

1.1.18 Morse functions. Let M be a manifold and let I be an interval. A *Morse function* is a smooth map $f : M \to I$ all of whose critical points are nondegenerate. We will mostly be concerned with manifolds with boundary; then we will always require that

$$f^{-1}(\partial I) = \partial M.$$

Finally we will require that the two boundary points of the interval I are regular values. In other words, there are no critical points on ∂M.

If M is compact (which will always be our case), a Morse function has only finitely many critical points. It is possible to arrange a Morse function such that these critical points all have distinct images in I. We will always assume this is the case.

1.1.19 Theorem. *(See Hirsch [27], 6.1.2.) Morse functions always exist. (In fact most functions are Morse functions, in the sense that in the space of all smooth maps $M \to I$ the Morse functions form a dense subset.)*

Exercises

1. Show that the unit interval $I = [0, 1]$ is indeed a manifold with boundary, by covering it with two half-lines H^1.
2. Make sense of the definition of in-boundary and out-boundary in the case of a manifold M without boundary, i.e. $\partial M = \varnothing$.
3. Discuss the validity of the arguments given on page 14, starting with *Hey, if you invert the order of the factors,...* The difficult case to understand is $X = Y$.
4. Let W be a complex vector space (of finite dimension over $\mathbb{C}$). Every vector $w \neq 0$ spans a vector space which is 2-dimensional over $\mathbb{R}$, with basis $[w, iw]$. Similarly, a complex ordered basis $[w_1, \ldots, w_n]$ for W defines a real ordered basis $[w_1, iw_1, \ldots, w_n, iw_n]$. Show that every $\mathbb{C}$-linear map $V \to W$ is orientation preserving over $\mathbb{R}$.
5. Let M be a closed manifold and consider smooth maps $f : M \to \mathbb{R}$. Put $H := \{x \in \mathbb{R} \mid x \geq 0\}$. One could expect that $f^{-1}(H)$ would be a manifold with boundary (the boundary being $f^{-1}(0)$). Give an example to show that this is not true in general, even if $f^{-1}(0)$ is nonempty. Under what conditions on f is it true?

6. Let B denote the interval $[-1, 1]$, and consider functions $B \to B$. Show that $x \mapsto x^2$ is a Morse function, while $x \mapsto x^3$ is not.

1.2 Cobordisms

Our real interest is in oriented cobordisms, but let us first have a look at the unoriented case.

Unoriented cobordisms

1.2.1 Unoriented cobordisms. Let there be given two closed manifolds (i.e. compact, without boundary) Σ_0 and Σ_1 of dimension $n - 1$. A *cobordism* between Σ_0 and Σ_1 is a compact n-manifold M whose boundary is $\Sigma_0 \coprod \Sigma_1$. When a cobordism exists, Σ_0 and Σ_1 are said to be *cobordant*.

(In order to admit cobordisms from a given manifold to itself, more precisely we need to define a cobordism as a certain map from $\Sigma_0 \coprod \Sigma_1$ onto the boundary of M. We will explain this technicality in a minute, when we come to oriented cobordisms.)

The prefix 'co-' in cobordism has nothing to do with duality as it is used in categorical language (as in the words 'coalgebra' and 'coproduct'). It simply means 'together', like in 'cooperation' or 'coproduction'. Originally, a single manifold Σ was called *bordant* if it formed the boundary of some manifold M; then two manifolds were called cobordant if together they formed the boundary of some manifold M.

1.2.2 Examples. Here are two examples ($n = 2$):

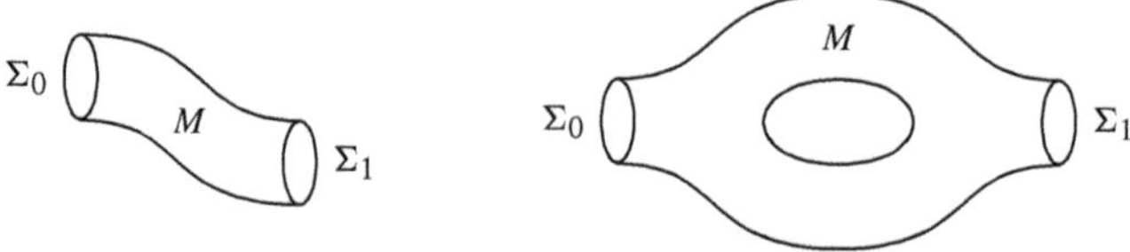

In these two examples, Σ_0 and Σ_1 are both connected. But we have not excluded the possibility that they be disconnected – in fact, disconnected manifolds are crucial for the theory – so here is an example of a cobordism between a single circle Σ_0 on one side and a pair of circles Σ_1 on the other side:

(1.2.3)

and here is a cobordism between the union of three circles on one side and the empty 1-manifold on the other side!

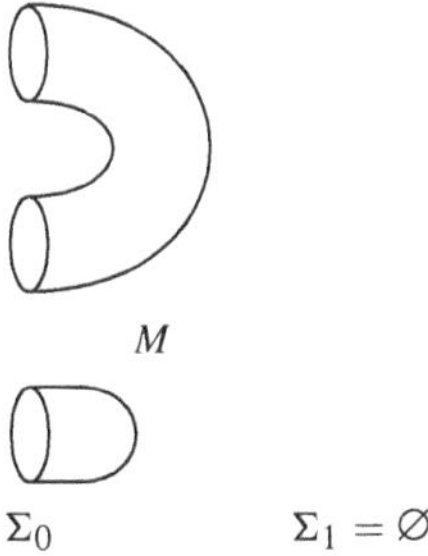

1.2.4 Movies and histories. A cobordism can be thought of as an interpolation between the two manifolds. Other analogies you can have in mind are the notion of morphing in computer graphics, or simply a history or a movie. Example 1.2.3 is the history of a single circle breaking up into two circles. Here are some frames of the movie:

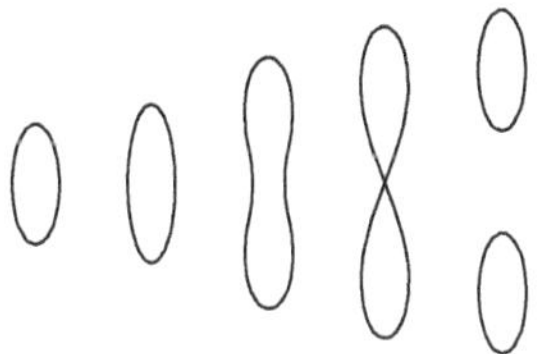

If you are a string theorist you can think of this as a string (elementary particle) which propagates in time, and degenerates into two strings, so this is a sort of thickening of the Feynman diagram

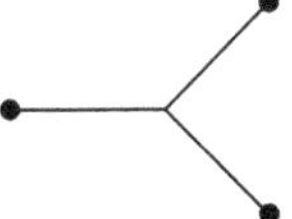

We think of the cobordism as describing an evolution in time, say from time $t = 0$ to time $t = 1$. So what is going on is that we consider a smooth map from M to the unit interval $I = [0, 1]$ (the time line) such that Σ_0 maps to 0 and Σ_1 maps to 1. This is the Morse theory viewpoint – the figure was drawn in 1.1.14. In all these examples we get the impression that we are just projecting down on the interval, but in general we regard our manifolds as abstract without embedding into any ambient pre-existing space-time, and without any canonical 'time line' interval to project down onto. There is no absolute time, and any function $M \to I$ (mapping Σ_0 to 0 and Σ_1 to 1) will do to define a sort of movie, a new viewpoint of the circle splitting up into two.

Because of the history analogy (which is quite relevant in view of the physical origin), the following two cobordisms are called the birth of a circle and the death of a circle, respectively:

(but here we are already anticipating the notion of oriented cobordism).

1.2.5 Example: a higher-dimensional cobordism. While these constructions are valid in any dimension, it is difficult to make illustrations in dimensions higher than 2. Just to stress that we are not talking only about circles and tubes, let us imagine a cobordism in dimension 3, from a torus to a sphere. Take a torus and start to pinch a certain circle on it, until this circle is just a point – this is the critical frame of the film – and continue: the two sides of the pinching circle break apart, and the whole surface becomes the skin of a sausage, which is then gradually contracted until it is a sphere. The movie viewpoint is convenient then, since each frame is of dimension 2, so we still have a fair chance of drawing pictures:

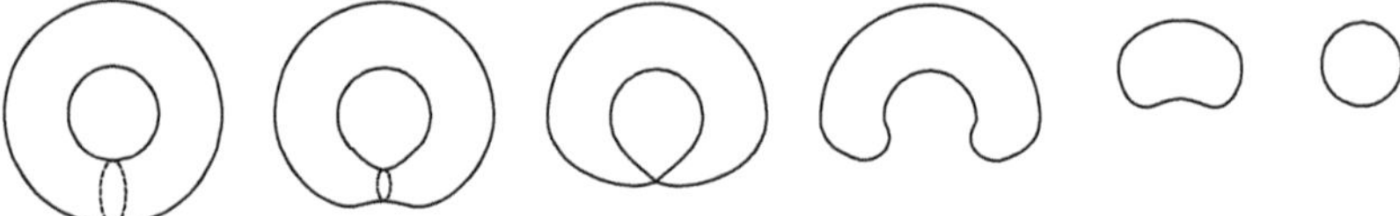

In this film, just as in the one in 1.2.4 there is a frame with a singular manifold – is that a problem? No, this is normal, and in fact necessary in order to make a transition between two manifolds which are not diffeomorphic (cf. the regular interval theorem quoted below in 1.3.8). The manifold M realising the cobordism can still be smooth.

1.2.6 Caution. It is important to note that our manifolds are not embedded in any ambient space – they are just abstract manifolds. Thus it has no meaning to talk about crossing over or under, or being entangled. For example, this drawing of a cobordism from a circle to a circle:

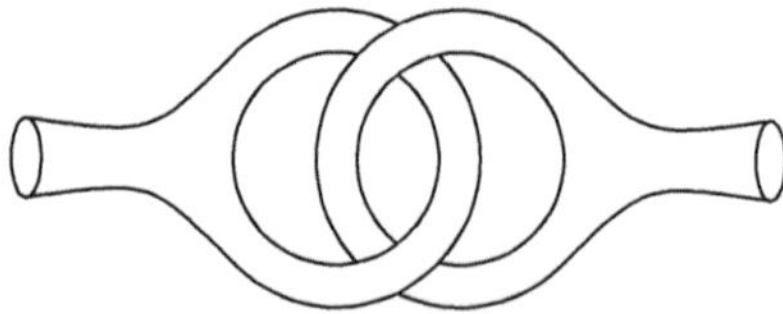

is merely a product of the artist's perversion. A more honest and less confusing drawing of the same cobordism would be this:

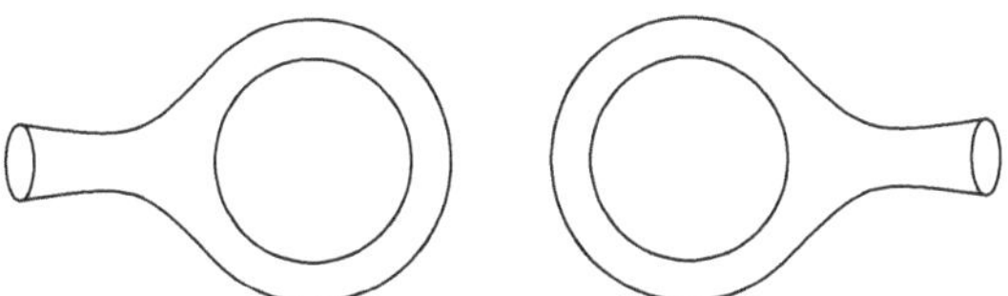

because as abstract manifolds they are the same. (What puzzles us when we see the conjurer perform this trick in the circus is that the whole circus is embedded in $\mathbb{R}^3$, where it ought to be impossible to separate the two connected components.)

1.2.7 Example: some 1-dimensional cobordisms. In this drawing,

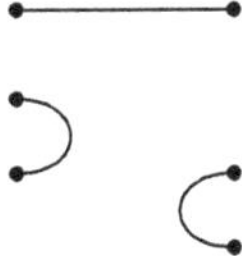

you can either think of three different cobordisms which are drawn close to each other (the first one between two points, the second and third go between a pair-of-points and the empty set), or you can think of it as one cobordism between a three-point manifold and another three-point manifold.

1.2.8 When are two $(n-1)$-manifolds cobordant? That is, given two closed manifolds Σ_0 and Σ_1, when does there exist a cobordism M between them? In general this question is difficult; we will take it up briefly in 1.2.19. In dimensions 0 and 1 it is not difficult to tell:

1.2.9 Lemma. *Two closed* 0*-manifolds are cobordant if and only if they have the same number of points modulo* 2.

Proof. Any pair of points can be joined by a curve (like in the previous figure above). On the other hand every 1-manifold with boundary has an even number of boundary components. So a cobordism between Σ_0 and Σ_1 exists if and only if there is an even number of points all together in $\Sigma_0 \coprod \Sigma_1$. □

1.2.10 Lemma. *Any two closed* 1*-manifolds are cobordant.*

Proof. A closed 1-manifold is a disjoint union of circles. So an easy way to construct a cobordism between a manifold **m** consisting of m circles and another manifold **n** consisting of n circles is to take m copies of 'death-of-a-circle' and n copies of 'birth-of-a-circle':

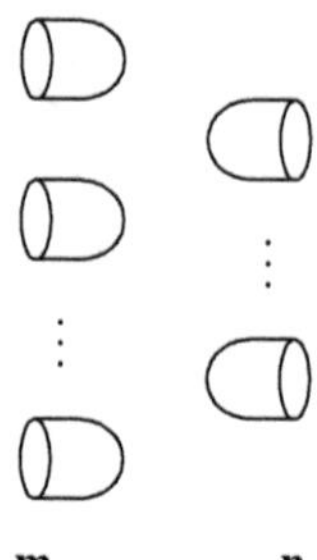

□

Oriented cobordisms

Now since we regard Σ_0 and Σ_1 as initial and final configurations and think of the cobordism M as describing a time evolution, it is natural we should want a clearer notion of direction – an arrow of time. Another reason is that we want to construct a *category* of cobordisms, so we need arrows, not just interpolations. To get a clear notion of the direction of the cobordism it is natural to use orientations.

1.2.11 Oriented cobordisms. Let Σ_0 and Σ_1 now be closed oriented $(n-1)$-manifolds. Intuitively, an oriented cobordism from Σ_0 to Σ_1 is an oriented n-manifold M whose in-boundary is Σ_0 and whose out-boundary is Σ_1. We will write it

$$\Sigma_0 \overset{M}{\Rightarrow} \Sigma_1.$$

(Note that in order to specify an oriented cobordism it is not enough to give an oriented manifold with boundary: we need also an orientation of the boundary, in order to be able to tell which boundary components are in-boundaries and which are out-boundaries.)

Now the above definition is still not good enough, because we would like very much to have cobordisms from a given Σ to itself, and this is not possible with the above definition, since a manifold cannot at the same time be an in-boundary and an out-boundary for a manifold M. What we need is a more relative description: instead of considering the source and target of the cobordism as submanifolds, we will just require them to be embedded in M. So here comes the final, official definition of a cobordism:

Let Σ_0 and Σ_1 be closed oriented $(n-1)$-manifolds. An *oriented cobordism* from Σ_0 to Σ_1 is an compact oriented manifold M together with smooth maps

$$\Sigma_0 \rightarrow M \leftarrow \Sigma_1$$

such that Σ_0 maps diffeomorphically (preserving orientation) onto the in-boundary of M, and Σ_1 maps diffeomorphically (preserving orientation) onto the out-boundary of M.

Note that this is practically the same as the first definition we gave, but it gives more flexibility. We can now have a a cobordism from Σ to Σ, provided we can find two diffeomorphisms: one from Σ to the in-boundary, and another from Σ to the out-boundary. (We will construct such maps in a minute.)

To begin with, we will often picture a cobordism by drawing small arrows on the boundary to represent a positive normal. Whenever possible, we will draw a cobordism with the in-boundary on the left and the out-boundary on the right, and in many cases this will allow us to dispense with drawing the small arrows . . .

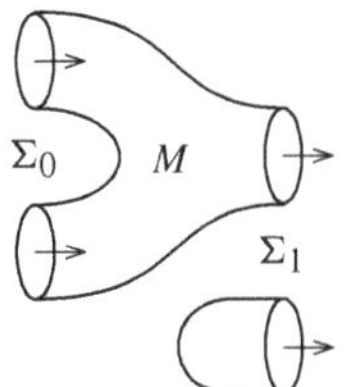

1.2.12 Cobordisms are not functions! An oriented cobordism is something that goes *from* one manifold Σ_0 *to* another manifold Σ_1. But it is worth stressing that it is not a function! It makes no sense to ask what it does to a particular point of Σ_0. (The most remarkable example is perhaps that we can have a cobordism from a nonempty manifold to $\emptyset$. This is not possible with functions of any kind . . .)

1.2.13 The closed interval. The most fundamental of all cobordisms is the closed interval. To be concrete, take the unit interval $I = [0, 1]$ with its standard orientation, and with the boundary points 0 and 1 given standard orientation $+$ as well. We already noted in 1.1.12 that 0 is an in-boundary and 1 is an out-boundary, so I defines a cobordism from 0 to 1.

There are obvious generalisations of this construction. Given any two (positively oriented) one-point manifolds p_0 and p_1, we can simply map p_0 to 0 and p_1 to 1; these two maps are clearly orientation-preserving diffeomorphisms onto the boundary of I:

$$p_0 \to I \leftarrow p_1.$$

In this way I provides a cobordism from p_0 to p_1. This illustrates the flexibility provided by the definition of cobordisms in terms of maps instead of submanifolds – one and the same manifold I can serve to construct cobordisms between various distinct objects, by using different embeddings onto its boundary.

In the same vein, we can replace the unit interval by any other simple oriented path M. Precisely, taking an orientation-preserving diffeomorphism $I \xrightarrow{\sim} M$ we get a new cobordism by composing with it:

$$\begin{array}{ccccc} & & M & & \\ & \nearrow & \uparrow & \nwarrow & \\ p_0 & \rightarrow & I & \leftarrow & p_1 \end{array}$$

Here is a drawing summarising these constructions:

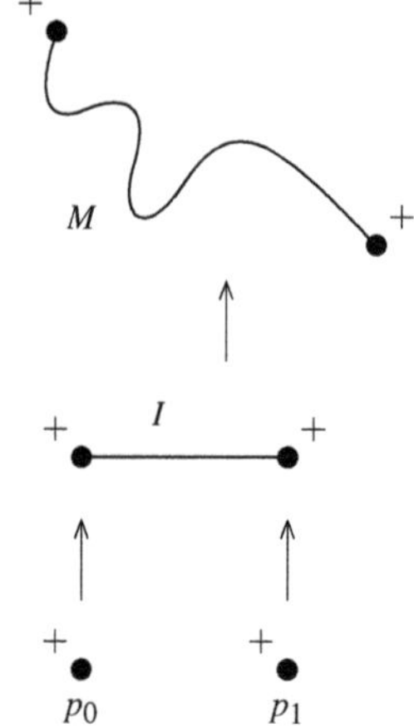

1.2.14 Example: other cobordisms between points. The example of the unit interval admits eight variations, by taking different orientations of the three manifolds involved. If we keep I with its standard orientation there are four possibilities. If 0 and 1 are both positively oriented we have

If we take 0 with orientation $+$, and 1 with orientation $-$, then both are in-boundaries! So we have produced a cobordism from a two-point manifold (with total sign zero) to the empty 0-manifold:

If both boundary points are oriented by a minus, then we get a cobordism from 1 to 0. Forgetting about the names of the manifolds we picture it like this:

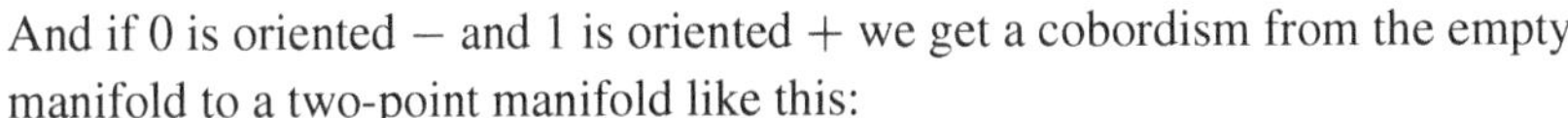

And if 0 is oriented $-$ and 1 is oriented $+$ we get a cobordism from the empty manifold to a two-point manifold like this:

$\varnothing$

(Taking reverse orientation of I we would get four other configurations but they would differ from the above four cases only by the names, which in any case is immaterial.)

1.2.15 Lemma. *Two oriented* 0*-manifolds* Σ_0 *and* Σ_1 *are cobordant if and only if the 'sum of the signs of* Σ_0*' equals the 'sum of the signs of* Σ_1*'.*

Here the 'sum of the signs' means this: suppose Σ_0 has n points with positive orientation and m points with negative, then the sum of its signs is $n - m$.

Proof. Exercise. (Use the drawings above.) □

1.2.16 Cylinders. The interval construction can be used to construct easy and important examples of cobordisms in higher dimensions: the cylinders. They are cobordisms from any given manifold to itself. To construct one, take a closed oriented manifold Σ and cross it with the unit interval I, with its standard orientation (and with standard orientation on the boundary points). The boundary of $\Sigma \times I$ consists of two copies of Σ: one which is an in-boundary, $\Sigma \times \{0\}$, and another which is an out-boundary, $\Sigma \times \{1\}$. So we get a cobordism from Σ to Σ by taking the obvious maps

$$\begin{aligned} \Sigma &\xrightarrow{\sim} \Sigma \times \{0\} \subset \Sigma \times I \\ \Sigma &\xrightarrow{\sim} \Sigma \times \{1\} \subset \Sigma \times I. \end{aligned}$$

As in the interval example, the same construction serves to give a cobordism between any pair of $(n-1)$-manifolds Σ_0 and Σ_1 both of which are diffeomorphic to Σ; just take

$$\begin{aligned} \Sigma_0 &\xrightarrow{\sim} \Sigma \xrightarrow{\sim} \Sigma \times \{0\} \subset \Sigma \times I \\ \Sigma_1 &\xrightarrow{\sim} \Sigma \xrightarrow{\sim} \Sigma \times \{1\} \subset \Sigma \times I. \end{aligned}$$

And again, any orientation-preserving diffeomorphism $\Sigma \times I \xrightarrow{\sim} M$ will also define a cobordism $M : \Sigma \Rightarrow \Sigma$, or if we combine the two variations, a cobordism $M : \Sigma_0 \Rightarrow \Sigma_1$:

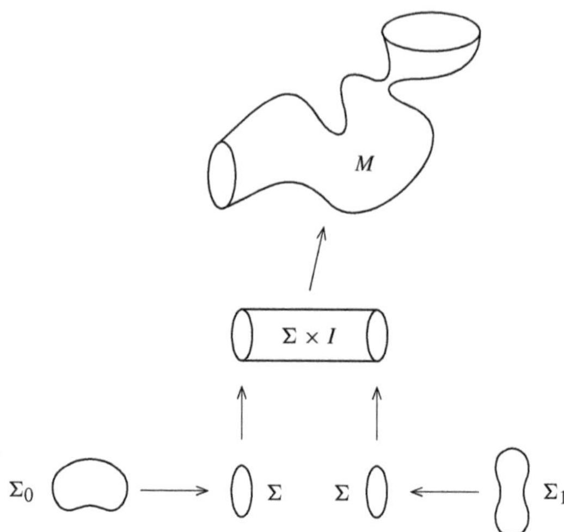

So in conclusion: for any two diffeomorphic manifolds Σ_0 and Σ_1 there exists a cobordism from Σ_0 to Σ_1, and in fact there are MANY! Those produced in this way are all *equivalent* cobordisms in the sense we now make precise.

1.2.17 Equivalent cobordisms. Given two oriented cobordisms from Σ_0 to Σ_1,

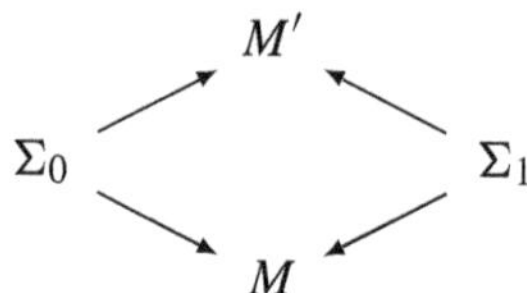

we say they are *equivalent* if there is an orientation-preserving diffeomorphism $\psi : M \xrightarrow{\sim} M'$ making this diagram commute:

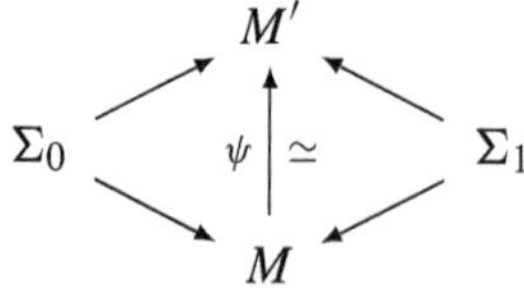

(Note that the two triangles truly commute – not just up to diffeomorphism. Thinking of Σ_0 and Σ_1 as submanifolds in M (and in M') this means that ψ induces the identity on the boundaries.) In the next subsection we will divide out by these equivalences, and consider equivalence classes of cobordisms, called *cobordism classes*.

1.2.18 'U-tubes'. In analogy with Example 1.2.14, we can consider cylinders with one of the boundaries reversed. Precisely, given a closed manifold Σ, map it onto one end of the cylinder $\Sigma \times I$, and map $\overline{\Sigma}$ onto the other end. (Recall that $\overline{\Sigma}$ means the same manifold but with opposite orientation.) Now both boundaries are in-boundaries (and the out-boundary is empty):

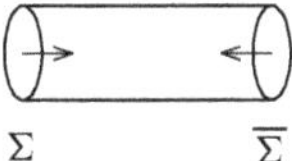

We will often draw such a cylinder like this:

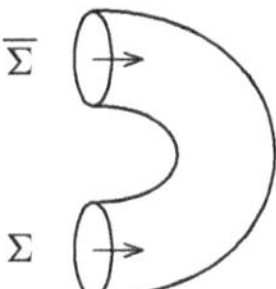

just to keep the convention of having in-boundaries on the left, and out-boundaries on the right, but the reader should be warned not to take the drawings too literally: the two figures are meant to represent the *same* cobordism, while our embedded-in-$\mathbb{R}^3$ intuition would rather suggest that the two cobordisms are distinct (but equivalent in the sense of 1.2.17).

Similarly we can form a cylinder with two out-boundaries

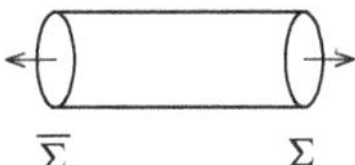

which we could draw like this:

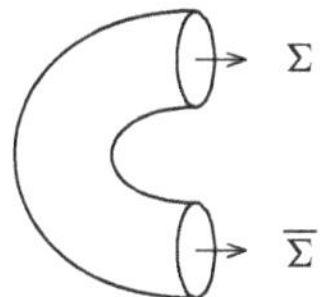

1.2.19 Digression on Thom's theory of cobordism groups. Cobordisms were introduced in the 1950s by L. S. Pontryagin and by R. Thom [47]. The relation that two closed oriented n-manifolds be (oriented) cobordant is an equivalence relation. The set of such equivalence classes of closed oriented n-manifolds is an abelian group Ω^n under disjoint union. (These equivalence classes of closed oriented manifolds must not be confused with the equivalence classes of cobordisms mentioned in 1.2.17.) We saw in 1.2.15 that the oriented cobordism group in dimension 0 is $\mathbb{Z}$, and the argument given in 1.2.10 shows that for $n = 1$ the group is 0: any two closed oriented 1-manifolds are cobordant. A famous result of Thom [47] states that

If the dimension n is not divisible by 4 *then the oriented cobordism group Ω^n is finite abelian. For $n = 4k$, the oriented cobordism group Ω^n is a finite*

abelian group times $\pi(k)$ copies of $\mathbb{Z}$, where $\pi(k)$ denotes the number of partitions of k. For low n the results are

$$\Omega^0 = \mathbb{Z}, \quad \Omega^1 = \Omega^2 = \Omega^3 = 0, \quad \Omega^4 = \mathbb{Z}, \quad \Omega^5 = \mathbb{Z}/2\mathbb{Z}, \quad \Omega^6 = \Omega^7 = 0.$$

The first nontrivial closed manifold which is not the boundary of any manifold is the complex projective plane $\mathbb{CP}^2$ (which generates Ω^4).

Decomposition of cobordisms

We have been led by the previous examples to think of cobordism as a sort of generalisation of intervals. Pursuing this analogy, we will now decompose cobordisms, cutting them in halves, just as one can subdivide an interval and get two shorter intervals.

1.2.20 Decomposition of cobordisms. An important feature of cobordisms is that you can *decompose* them. In the movie analogy, this means that we take some intermediate frame (corresponding to time t) and regard it as a submanifold in M which splits M into two parts (not necessarily connected). Precisely, take a smooth submanifold Σ_t which divides M into two parts, with all the in-boundaries in one part and all the out-boundaries in the other part. Give Σ_t orientation such that its positive normal points towards the out-part. The nicest way of arranging such a cut is to take a smooth map $f : M \to [0, 1]$ such that $f^{-1}(0) = \Sigma_0$ and $f^{-1}(1) = \Sigma_1$, and make the cut along the inverse image of a regular value t, oriented such that the positive normal points towards the out-boundaries, just as the positive normal of $t \in [0, 1]$ points towards 1. Then we have a picture like this:

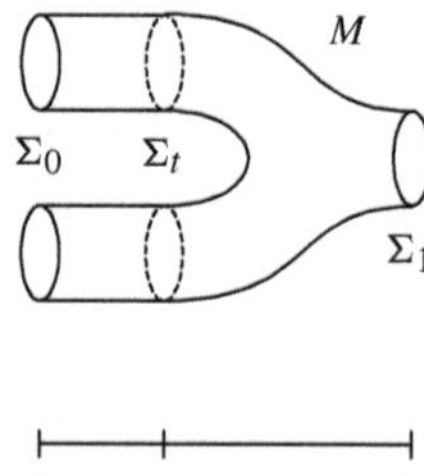

The result is two new cobordisms: one from Σ_0 to Σ_t given by the piece $M_{[0,t]} := f^{-1}([0, t])$, and another from Σ_t to Σ_1 given by the piece $M_{[t,1]} := f^{-1}([t, 1])$.

(Note that this operation is something that has no analogue for functions or arrows in a concrete category: you cannot just take an arbitrary function and look at its value halfway to the target!)

In the next section we will reverse this process and show how to compose two cobordisms, provided they have compatible boundaries.

1.2.21 'Snake decomposition' of a cylinder. Starting with a cylinder $C = \Sigma \times I$ over some closed manifold Σ we could of course decompose it by cutting it in the middle like this:

Σ Σ Σ

getting a decomposition of the cylinder into two cylinders.

But we could also reverse the orientation of the middle copy of Σ:

Σ $\overline{\Sigma}$ Σ

This is not a decomposition in the sense we have just described because the positive normal of the middle $\overline{\Sigma}$ does not point towards the out-boundary of the original cobordism. But if we cut a little bit more we can repair that defect:

C_0 U_1 U_0 C_1

Σ Σ $\overline{\Sigma}$ Σ Σ

This is a true decomposition, provided we interpret the three pieces in the correct way. The in-part of the decomposition is $M_0 := C_0 \coprod U_0 : \Sigma \Rightarrow \Sigma \coprod \overline{\Sigma} \coprod \Sigma$. The out-part of the decomposition is $M_1 := U_1 \coprod C_1 : \Sigma \coprod \overline{\Sigma} \coprod \Sigma \Rightarrow \Sigma$.

If we draw the pieces U_0 and U_1 as U-tubes as in 1.2.18 it is easier to grasp the decomposition:

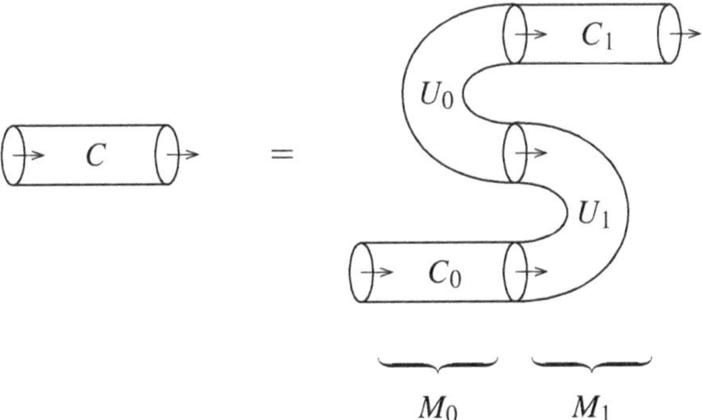

In conclusion we have found a decomposition of a cylinder into two cobordisms which are not cylinders. This particular example will have important consequences for TQFTs...

1.2.22 Disconnectedness and disjoint union. We have already seen many examples of disconnected cobordisms. The example above illustrates the true importance of allowing disconnected objects (manifolds and cobordisms): namely that even if we start with a connected cobordism, when we chop it up we easily get disconnected ones! This is good, because if we have a disconnected cobordism we can study its connected components separately – and the more we can split up the problem into simpler parts the more easily we can understand it. In general, a disconnected manifold can always be written as the disjoint union of its connected components. With a little care with the boundaries, it is almost as easy for cobordisms: in the example above we chopped up a cobordism, and without really worrying about it we arranged the resulting pieces in such a way that we could interpret them as disjoint unions of simpler pieces. . .

Topological quantum field theories

Roughly, a quantum field theory takes as input *spaces* and *space-times* and associates to them *state spaces* and *time evolution operators*. The space is modelled as a closed oriented $(n-1)$-manifold, while space-time is an oriented n-manifold whose boundary represents time 0 and time 1. The state space is a vector space (over some ground field $\Bbbk$), and the time evolution operator is simply a linear map from the state space of time 0 to the state space of time 1. The theory is called *topological* if it only depends on the topology of the space-time. This means that 'nothing happens' as long as time evolves cylindrically. . .

Mathematical axioms for TQFTs were put forth in the late 1980s: first Segal [45] proposed a set of axioms for the related notion of *conformal* quantum field theories, and shortly after, Atiyah [5] gave a set of axioms for a topological quantum field theory. The following is a slight rewrite of his axioms.

1.2.23 Topological quantum field theories. An *n-dimensional topological quantum field theory* (TQFT) is a rule $\mathscr{A}$ which to each closed oriented $(n-1)$-manifold Σ associates a vector space $\Sigma\mathscr{A}$, and to each oriented cobordism $M : \Sigma_0 \Rightarrow \Sigma_1$ associates a linear map $M\mathscr{A}$ from $\Sigma_0\mathscr{A}$ to $\Sigma_1\mathscr{A}$.

This rule $\mathscr{A}$ must satisfy the following five axioms.

A1: Two equivalent cobordisms must have the same image:

$$M \cong M' \Rightarrow M\mathscr{A} = M'\mathscr{A}.$$

A2: The cylinder $\Sigma \times I$, thought of as a cobordism from Σ to itself, must be sent to the identity map of $\Sigma\mathscr{A}$.

A3: Given a decomposition $M = M'M''$ then

$$M\mathscr{A} = (M'\mathscr{A})(M''\mathscr{A}) \quad \text{(composition of linear maps).}$$

A4: Disjoint union goes to tensor product: if $\Sigma = \Sigma' \coprod \Sigma''$ then $\Sigma\mathscr{A} = \Sigma'\mathscr{A} \otimes \Sigma''\mathscr{A}$. This must also hold for cobordisms: if $M : \Sigma_0 \Rightarrow \Sigma_1$ is the disjoint union of $M' : \Sigma'_0 \Rightarrow \Sigma'_1$ and $M'' : \Sigma''_0 \Rightarrow \Sigma''_1$ then $M\mathscr{A} = M'\mathscr{A} \otimes M''\mathscr{A}$.

A5: The empty manifold $\Sigma = \emptyset$ must be sent to the ground field $\Bbbk$. (It follows that the empty cobordism (which is the cylinder over $\Sigma = \emptyset$) is sent to the identity map of $\Bbbk$.)

The first two axioms express that the theory is topological: the evolution depends only on the diffeomorphism class of space-time, not on any additional structure like metric or curvature...

Axiom A4 reflects a standard principle of quantum mechanics: that the state space of two independent systems is the tensor product of the two state spaces. (Axiom A5 also reflects this principle.)

1.2.24 Towards a categorical interpretation of the axioms. The first three axioms will amount to saying that the rule $\mathscr{A}$ is a *functor* – for this to make sense we must of course specify in which sense manifolds and cobordisms form a category. This is the subject of the next section.

Axioms A4 and A5 in turn amount to saying that this functor is furthermore *monoidal*. Monoidal categories and functors are what Chapter 3 is about, but we will anticipate the definition: roughly a monoidal category is one equipped with a 'multiplication' with neutral object. In our case, for manifolds and cobordisms the 'multiplication' is disjoint union, and the neutral object for that operation is the empty manifold. For vector spaces, the 'multiplication' is the tensor product, and the neutral object is the ground field. A monoidal functor is one that preserves such monoidal structure.

We will have a lot more to say about disjoint union in 1.3.24.

Let us see how the axioms work, and extract some important consequences of them. The next couple of arguments depend on some linear algebra (pairings and copairings) which is carefully explained in Section 2.1.

1.2.25 Nondegenerate pairings and finite-dimensionality. Take any closed manifold Σ, let $V := \Sigma\mathscr{A}$ be its image under a TQFT $\mathscr{A}$, and let W denote the image of $\overline{\Sigma}$. The image of the U-tube of Example 1.2.18 is then a pairing $\beta : V \otimes W \to \Bbbk$. Similarly, the other U-tube is sent to a copairing $\gamma : \Bbbk \to W \otimes V$ (see 2.1.10). Now consider the snake decomposition of the cylinder $\Sigma \times I$ described in 1.2.21. The axioms imply that the composition of linear maps

$$V \xrightarrow{\mathrm{id}_V \otimes \gamma} V \otimes W \otimes V \xrightarrow{\beta \otimes \mathrm{id}_V} V \tag{1.2.26}$$

is the identity map. Indeed, A4 says that the two parts of the snake decomposition have these images under $\mathscr{A}$:

$$\mapsto \quad \mathrm{id}_V \otimes \gamma \qquad\qquad \mapsto \quad \beta \otimes \mathrm{id}_V .$$

Axiom A3 implies that the composition of these two maps is the image of the original cylinder, which by A2 is the identity map of V.

Now the fact that the map 1.2.26 is the identity map (together with the equation obtained from a snake going the other way), is precisely to say that the pairing β is nondegenerate (cf. Definition 2.1.10). So we have shown this:

1.2.27 Proposition. *Let $\mathscr{A}$ be a TQFT. The image vector space V of a closed manifold Σ comes equipped with a nondegenerate pairing with $W := \overline{\Sigma}\mathscr{A}$.*

Given such a nondegenerate pairing there is a canonical identification of W with V^* the dual space of V. In Atiyah's original formulation, this is an axiom:

$$\Sigma \mapsto V \quad \Rightarrow \quad \overline{\Sigma} \mapsto V^* .$$

The notion of nondegeneracy is a strong one, which in fact implies (cf. 2.1.12)

1.2.28 Corollary. *The image vector spaces in a TQFT are necessarily of finite dimension.*

The proposition illustrates how topological properties on the manifold side of a TQFT translate into algebraic structure on the vector space side. This particular result does not depend on the dimension n of the theory. In the case $n = 2$, there is even more structure on the vector spaces: they turn out to be Frobenius algebras (cf. Theorem 3.3.2).

1.2.29 Example. From a mathematical point of view, the importance of TQFTs is that they produce invariants of manifolds: suppose M is an n-manifold without boundary, then it can be considered as a cobordism $\emptyset_{n-1} \Rightarrow \emptyset_{n-1}$, so $\mathscr{A}$ associates to it a linear map $\Bbbk \to \Bbbk$, i.e. a constant, which is a topological invariant of the manifold. Furthermore, this invariant can be computed by cutting up M (in many ways). It is beyond the scope of these notes to give any significant examples of this (the reader is referred to Atiyah [6]), but let us briefly mention a very simple example.

Let $\mathscr{A}$ be a TQFT. Take a closed manifold Σ, and let V be the image vector space $\Sigma\mathscr{A}$. Consider the manifold $M = \Sigma \times S^1$ (which is without boundary) and decompose it like this:

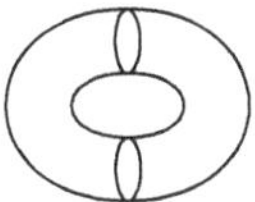

The cut is the disjoint union $\Sigma \coprod \overline{\Sigma}$, and the two U-tubes are a sent to a copairing γ and a pairing β on V, as we just explained in 1.2.25. So the invariant $M\mathscr{A}$ is the composite $\gamma\beta : \mathbb{k} \to \mathbb{k}$. Now a little linear algebra (Exercise 10 below) shows that this is nothing but the trace of the identity map of V, so the invariant associated to M is $\dim V$.

Exercises

1. Let M denote a long-playing record oriented such that the in-boundary is the circumference Σ_0, and the out-boundary is the hole in the middle, Σ_1. Then the 2-manifold M realises a cobordism from Σ_0 to Σ_1. The exercise is to turn this static picture into a movie in the spirit of 1.2.4. (Hint: play the record...)
2. Let Σ_1 denote the surface of a solid doughnut M made of wood. Inside M there is a ball-shaped cavity whose wall we denote Σ_0. Arrange orientations such that Σ_0 is the in-boundary of M and Σ_1 is the out-boundary. So the 3-manifold M realises a cobordism from the sphere Σ_0 to the torus Σ_1. The exercise consists in turning this static picture into a movie in the style of 1.2.5.
3. Write down the proof of Lemma 1.2.15.
4. Draw examples of cobordisms from the empty 1-manifold $\emptyset_1$ to itself. Classify them all up to equivalence.
5. Show that the cobordism group described in 1.2.19 is indeed a group. (Hint: use 1.2.18.)

$\star$6. Use Thom's theorem (1.2.19) and the statement made there about $\mathbb{CP}^2$ to prove that there is no orientation-preserving diffeomorphism from $\mathbb{CP}^2$ to $\overline{\mathbb{CP}}^2$. In other words, there is no orientation-reversing diffeomorphism from $\mathbb{CP}^2$ to itself. (Hint: use 1.2.18.)

7. Consider a TQFT in dimension 2 with the following properties. The circle is sent to the vector space V of all n-by-n matrices over $\mathbb{k}$. For simplicity

take $n = 2$, but this has nothing to do with the dimension of the TQFT. By the axioms, $\Sigma \coprod \Sigma \mapsto V \otimes V$, etc., and $\emptyset_1 \mapsto \Bbbk$. Assume that is sent to the trace map

$$\begin{aligned} \varepsilon : V &\longrightarrow \Bbbk \\ \left(\begin{smallmatrix} a & b \\ c & d \end{smallmatrix}\right) &\longmapsto a + d, \end{aligned}$$

and that is sent to the multiplication map $\mu : V \otimes V \to V$ (matrix multiplication). Use the decomposition axiom A3 to determine the image of . (It must be a pairing $\beta : V \otimes V \to \Bbbk$.)

$\star$8. Continuing the previous exercise, assume that is sent to the map

$$\begin{aligned} \gamma : \Bbbk &\longrightarrow V \otimes V \\ 1 &\longmapsto \left(\begin{smallmatrix} 1 & 0 \\ 0 & 0 \end{smallmatrix}\right) \otimes \left(\begin{smallmatrix} 1 & 0 \\ 0 & 0 \end{smallmatrix}\right) + \left(\begin{smallmatrix} 0 & 0 \\ 1 & 0 \end{smallmatrix}\right) \otimes \left(\begin{smallmatrix} 0 & 1 \\ 0 & 0 \end{smallmatrix}\right) + \left(\begin{smallmatrix} 0 & 1 \\ 0 & 0 \end{smallmatrix}\right) \otimes \left(\begin{smallmatrix} 0 & 0 \\ 1 & 0 \end{smallmatrix}\right) \\ &\qquad + \left(\begin{smallmatrix} 0 & 0 \\ 0 & 1 \end{smallmatrix}\right) \otimes \left(\begin{smallmatrix} 0 & 0 \\ 0 & 1 \end{smallmatrix}\right). \end{aligned}$$

Show that this is not in contradiction with the snake decomposition of the cylinder, cf. 1.2.21 and 1.2.26. Precisely, show that γ is a copairing corresponding to the pairing β. Copairings (cf. 2.1.10) are always harder to understand than pairings, and messier to write down – there is usually no way to avoid long sums of tensors like this.

9. Continuing the previous two exercises, compute explicitly the composite $\gamma\beta : \Bbbk \to \Bbbk$. (According to Example 1.2.29 this should be multiplication by $4 = \dim V$.)

10. Verify this statement made in 1.2.29: if (β_{ij}) is an invertible symmetric n-by-n matrix and (γ^{ij}) is its inverse, then we have

$$\sum_{i,j} \gamma^{ij} \beta_{ij} = n.$$

(Hint: sum over one index at a time.)

1.3 The category of cobordism classes

We will now assemble cobordisms into a category: the objects should be closed oriented $(n-1)$-manifolds, and the arrows ought to be oriented cobordisms (but it turns out we need to consider cobordism *classes* instead). So we need to show how to compose two cobordisms (and check associativity), and we need to find an identity arrow for each object. Clearly we want the following property: if M is a cobordism and we decompose it into two parts M_0 and M_1 as in 1.2.20, then the composition of M_0 and M_1 should be M.

Intuitively, given one cobordism $M_0 : \Sigma_0 \Rightarrow \Sigma_1$ and another $M_1 : \Sigma_1 \Rightarrow \Sigma_2$ then the composite $M_0 M_1 : \Sigma_0 \Rightarrow \Sigma_2$ should be obtained by gluing together the manifolds M_0 and M_1 along Σ_1. This is a manifold with in-boundary Σ_0 and out-boundary Σ_2, and Σ_1 sits inside it as a submanifold

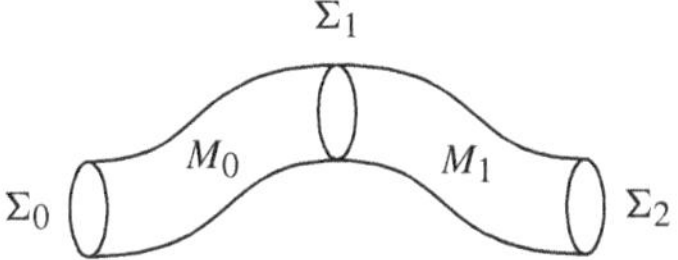

If you ask a differential topologist, he will typically tell you that this operation is well defined. But if you question him more closely, he will admit that the operation is only well defined up to diffeomorphism – but after all that is what matters in differential topology. If you press him really hard, it will come to light that $M_0 M_1$ is well defined up to diffeomorphism, but not up to *unique* diffeomorphism – in other words, there is no universal property... In order to appreciate these subtleties – and in any case to understand the statements – we will spend some time working out a simple example. From the viewpoint of everyday differential topology, this section is rather pedantic, but seen from a broader perspective it is interesting to be aware of such well-defined-up-to-something business.

Concerning identity arrows, the identity ought to be a cylinder of height zero, but such a 'cylinder' is not an n-manifold!

Both problems are solved by passing to diffeomorphism classes of cobordisms. Precisely, we identify cobordisms which are equivalent in the sense of 1.2.17 and let the arrows of our category be these equivalence classes, called *cobordism classes*.

Once we have specified the category structure we can also make more precise the notion of paralleling, which we have already touched upon (e.g. 1.2.22). It amounts to saying that our category is a monoidal category. At this point we will not need the precise definition; we just need to observe that there is a way to 'compose' cobordisms in parallel...

Gluing and composition

1.3.1 Gluing topological spaces. Let $f_0 : \Sigma \to M_0$ and $f_1 : \Sigma \to M_1$ be continuous maps between topological spaces. For simplicity we assume that M_0 and M_1 are disjoint (the problems related to the general case will be discussed in 1.3.24), and that the maps are injective. Now $M_0 \coprod_\Sigma M_1$ is defined by taking the (disjoint) union of M_0 and M_1 and quotienting by this equivalence relation: two points $m_0 \in M_0$ and $m_1 \in M_1$ are equivalent if there exists a point $x \in \Sigma$ such that $xf_0 = m_0$ and $xf_1 = m_1$. There are two natural maps $M_0 \to M_0 \coprod_\Sigma M_1 \leftarrow M_1$. The topology of $M_0 \coprod_\Sigma M_1$ is defined by declaring a subset open if its inverse images in M_0 and M_1 are both open. Intuitively $M_0 \coprod_\Sigma M_1$ is obtained by gluing M_0 and M_1 along their 'common' locus Σ.

We have this commutative diagram of continuous maps:

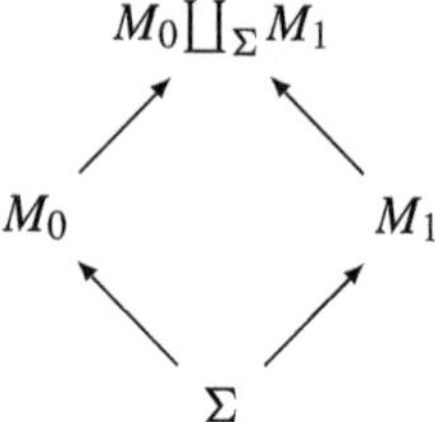

The crucial observation is that $M_0 \coprod_\Sigma M_1$, together with the diagram, is universal among all such commutative diagrams. For every commutative diagram

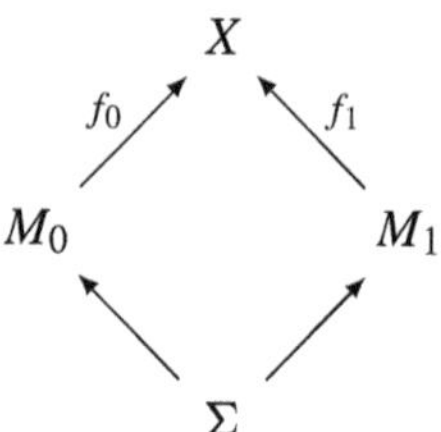

there exists a unique continuous map $f : M_0 \coprod_\Sigma M_1 \to X$ extending f_0 and f_1:

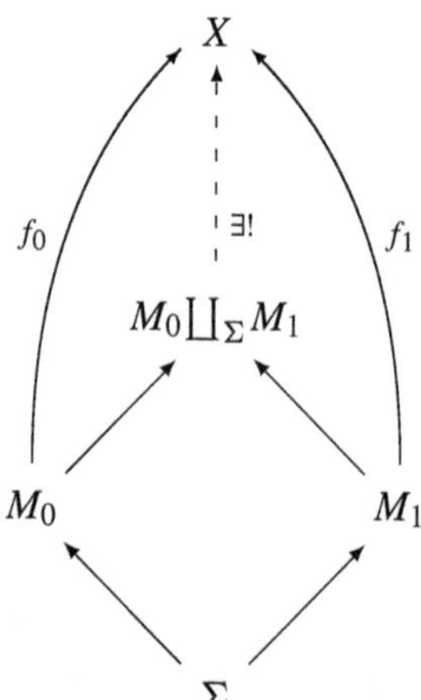

You can say loosely that the diagram of $M_0 \coprod_\Sigma M_1$ is the square nearest $M_0 \leftarrow \Sigma \rightarrow M_1$. The precise categorical statement is that $M_0 \coprod_\Sigma M_1$ is the *pushout*, or the *colimit* of $M_0 \leftarrow \Sigma \rightarrow M_1$ (cf. A.3.7 in the Appendix).

In the situation of the arbitrary diagram, the unique map $f : M_0 \coprod_\Sigma M_1 \rightarrow X$ can be thought of as the gluing of the two maps f_0 and f_1. So the statement can also be formulated like this: continuous maps glue if they agree on the border between their domains.

In the next paragraph, we use this to glue C^0 transition functions, and see that the pushout is also well defined for topological manifolds (i.e. topological spaces equipped with atlases).

1.3.2 Gluing topological manifolds. Suppose now M_0 and M_1 are topological manifolds with a common boundary component Σ. (As usual, what we really mean is that we have maps $M_0 \leftarrow \Sigma \rightarrow M_1$, from Σ onto the boundary components in question.) The question is whether the topological space $M_0 \coprod_\Sigma M_1$ is again a manifold in any canonical way. In other words, given the (maximal) C^0 atlases on M_0 and M_1, can we construct a C^0 atlas on $M_0 \coprod_\Sigma M_1$? Each point of $M_0 \coprod_\Sigma M_1$ which is not on the gluing locus Σ is already covered by a chart. Let us construct a chart $U \rightarrow \mathbb{R}^n$ around a point on Σ. By definition of the topology, the two restrictions $U_0 := U \cap M_0$ and $U_1 := U \cap M_1$ are open in M_0 and M_1, respectively. Observe that $U = U_0 \coprod_\Sigma U_1$ (where Σ now means $\Sigma \cap U$). By shrinking U if necessary, we can assume U_0 and U_1 are each domains for a chart, and since the atlases are maximal, we may as well assume that these charts are $f_0 : U_0 \rightarrow \mathbb{R}^n_- := \{x \in \mathbb{R}^n \mid x_n \leq 0\}$ and $f_1 : U_1 \rightarrow \mathbb{R}^n_+ := \{x \in \mathbb{R}^n \mid x_n \geq 0\}$. Now we already have the gluing (pushout) of these two half-spaces: together they form $\mathbb{R}^n$. So we have the solid diagram

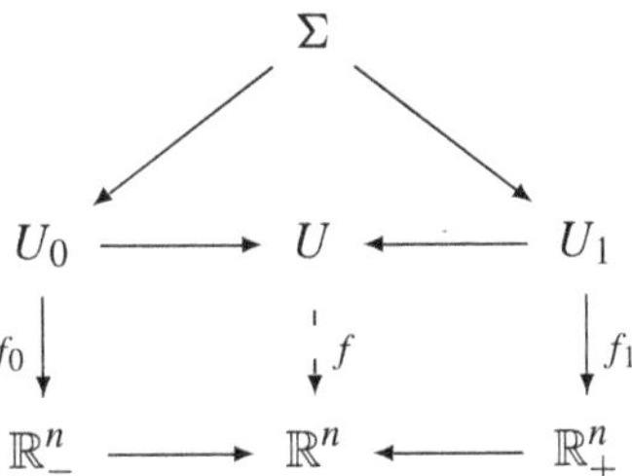

We need to construct the dashed map $f : U \rightarrow \mathbb{R}^n$, and show it is a homeomorphism. But this is all guaranteed by the universal property. First, the universal property of $U = U_0 \coprod_\Sigma U_1$ (with $\mathbb{R}^n$ in the place of X) implies there is a unique continuous map $U \rightarrow \mathbb{R}^n$ making the diagram commute. Second, the universal property of $\mathbb{R}^n$ provides a continuous map in the other direction, and clearly they are inverses to each other, so f is indeed a homeomorphism.

So we have constructed a coordinate chart with domain U. Now there were choices involved: for each choice of f_0 and f_1, the construction gives a chart f on U. We claim that all these charts have C^0 transition, so they belong to the same maximal atlas. Indeed, any other chart on U_1, say $g_1 : U_1 \to \mathbb{R}^n_+$, is related to f_1 by a continuous transition function $\alpha_1 : \mathbb{R}^n_+ \xrightarrow{\sim} \mathbb{R}^n_+$ (or possibly just open subsets of $\mathbb{R}^n_+$), and similarly for charts on U_0:

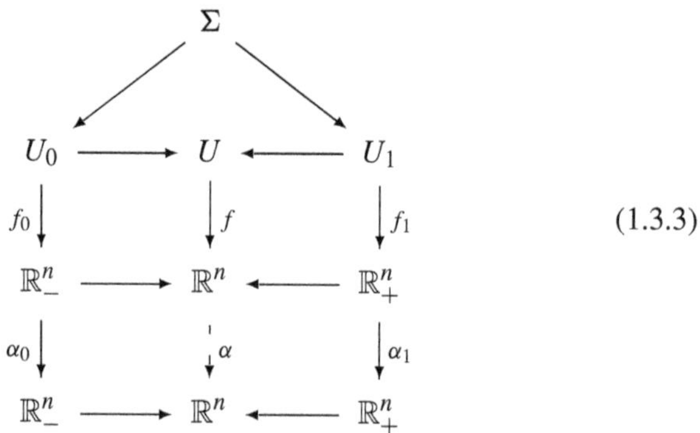

(1.3.3)

As for the f, the two charts g_0 and g_1 glue together to give a chart $g : U \to \mathbb{R}^n$. Now the coordinate change function for the two charts f and g on U is induced exactly by the coordinate changes on the half-charts. That is, $\alpha : \mathbb{R}^n \to \mathbb{R}^n$ is obtained by gluing α_0 and α_1. By the universal property, α is continuous.

So in conclusion, all the charts we construct belong to the same maximal atlas. So this defines a C^0 structure on $M_0 \coprod_\Sigma M_1$.

The reason for spending time with the continuous case is that we see exactly what will go wrong in the smooth case: smooth functions do not glue just because they take the same value on the border between their domains. As an easy concrete example:

1.3.4 Two distinct gluings of 1-manifolds. . . Given two smooth 1-manifolds M_0 and M_1, with a common boundary point p, we want to glue them. (You can think of the two intervals $[0, 1]$ and $[1, 2]$ which glue to give $[0, 2]$, if you wish.) We place ourselves in the situation of Diagram 1.3.3. Fix a chart on M_0 near p, say $f_0 : U_0 \to \mathbb{R}_-$. Now let us take two charts around p in M_1 and analyse their compatibility. We assume both charts have the same open set U_1 as domain. Start with $f_1 : U_1 \to \mathbb{R}_+$, and let $g_1 : U_1 \to \mathbb{R}_+$ be defined by composing with the coordinate change function

$$\begin{aligned} \alpha_1 : \mathbb{R}_+ &\longrightarrow \mathbb{R}_+ \\ x &\longmapsto x^2 \end{aligned}$$

(with inverse $y \mapsto \sqrt{y}$, both are smooth maps). So we have $g_1 = f_1\alpha_1$, and α_1 is the coordinate change function. So now we have two charts $U_1 \rightrightarrows \mathbb{R}_+$, neither of which can be said to be better or more canonical than the other.

Each of these charts glues with the chart on U_0 to give an $\mathbb{R}$-chart around p. We get one chart $f : U \to \mathbb{R}$ by gluing f_0 and f_1, and another chart $g : U \to \mathbb{R}$ by gluing $g_0 = f_0$ and g_1. Now it is easy to see that the transition between f and g is not smooth: the transition function α between these two charts of U is obtained by gluing $\alpha_0 = \mathrm{id}_{\mathbb{R}_-}$ with α_1, which gives

$$\begin{aligned} \alpha : \mathbb{R} &\longrightarrow \mathbb{R} \\ x &\longmapsto \begin{cases} x & \text{for } x \leq 0 \\ x^2 & \text{for } x \geq 0. \end{cases} \end{aligned}$$

So our example shows that there are at least two distinct maximal atlases on U which agree with the atlases on U_0 and U_1. In other words, we have two smooth structures on U, denoted (U, f) and (U, g).

1.3.5 ... which are diffeomorphic. Now, while these two structures are not identical, they may well be isomorphic – in fact we know they must be, since they are both connected 1-manifolds. To see it explicitly in this example, define $(U, f) \to (U, g)$ by taking

$$U \xrightarrow{f} \mathbb{R} \xrightarrow{g^{-1}} U.$$

We must check that this map is smooth and has a smooth inverse. Both these claims follow from the description of the map in terms of the local coordinates: it is simply the identity map $\mathbb{R} \to \mathbb{R}$! Caution: the description in local coordinates happens to be the identity map, but if you write out what happens to the points in U you see that the map itself is not the identity map. For the same reason, this diffeomorphism is not compatible with the inclusions $U_0 \to U \leftarrow U_1$, so there is no universal property.

Now we claim, furthermore, that every possible chart on U compatible with the charts on the half-parts is obtained from the above example via some coordinate change on the half-parts, and whichever be the obstruction for the resulting charts to belong to the same maximal atlas, it is clear that the above construction works also to establish a diffeomorphism between the two structures.

1.3.6 Smooth manifolds. Since we cannot get unique gluing (or even unique up to unique diffeomorphism), we should now head for showing that *some* gluing at least exists.

To find one smooth structure, we can replace our manifolds with diffeomorphic ones and try to glue them instead. If we succeed we can use the diffeomorphism to produce a smooth structure on the original manifold. In other words, finding a smooth structure on a topological manifold $M = M_0 M_1$ is equivalent to finding a smooth manifold S and a homeomorphism $M \stackrel{\sim}{\to} S$ which is a diffeomorphism on the two pieces M_0 and M_1. Given such a map, we can pull back the maximal atlas from S to get one on M.

1.3.7 Gluing of cylinders. As an example of this principle, we will now glue cylinders, and more generally cobordisms equivalent to cylinders. To glue two cylinders, say $\Sigma \times [0, 1]$ and $\Sigma \times [1, 2]$, we essentially do as in 1.3.4: take $\Sigma \times [0, 2]$. The smooth structure on $[0, 2]$ is any one constructed as in the example above – obviously we get a smooth structure on $\Sigma \times [0, 2]$ compatible with what we had on the two parts.

The next case to consider is how to compose two cobordisms $M_0 : \Sigma_0 \Rightarrow \Sigma_1$ and $M_1 : \Sigma_1 \Rightarrow \Sigma_2$ which are both equivalent to cylinders (in the sense of 1.2.17), say

$$\phi_0 : M_0 \stackrel{\sim}{\to} \Sigma_1 \times [0, 1]$$
$$\phi_1 : M_1 \stackrel{\sim}{\to} \Sigma_1 \times [1, 2].$$

We need to find a smooth manifold S and a homeomorphism $\phi : M_0 \coprod_{\Sigma_1} M_1 \to S$ whose restrictions to M_0 and M_1 are diffeomorphisms. But this is easy: just take

$$\phi := \phi_0 \coprod_{\Sigma_1} \phi_1 : M_0 \coprod_{\Sigma_1} M_1 \longrightarrow S := \Sigma_1 \times [0, 2]$$

defined by gluing ϕ_0 and ϕ_1; it is the map given by the universal property of the gluing in the category of continuous maps

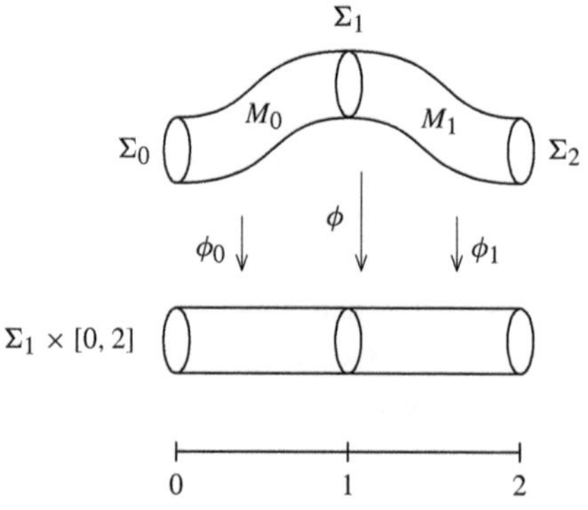

So ϕ is defined just as a homeomorphism. Now S has smooth structure which agrees with $\Sigma_1 \times [0, 1]$ and $\Sigma_1 \times [1, 2]$, and thus we get smooth structure on M via pullback along ϕ.

All this fiddling around with cylinders becomes important in view of the following result.

1.3.8 Regular interval theorem. (See Hirsch [27], 6.2.2.) *Let $M : \Sigma_0 \Rightarrow \Sigma_1$ be a cobordism and let $f : M \to [0, 1]$ be a smooth map without any critical points at all, and such that $\Sigma_0 = f^{-1}(0)$ and $\Sigma_1 = f^{-1}(1)$. Then there is a diffeomorphism from the cylinder $\Sigma_0 \times [0, 1]$ to M compatible with the projection to $[0, 1]$ like this:*

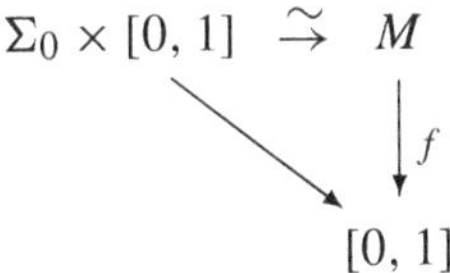

(And similarly, there is another diffeomorphism $\Sigma_1 \times [0, 1] \xrightarrow{\sim} M$ compatible with the projection.) □

1.3.9 Remark. In particular, this diffeomorphism induces a diffeomorphism $\psi : \Sigma_0 \xrightarrow{\sim} \Sigma_1$; this is the restriction to the out-boundary of $\Sigma_0 \times I \xrightarrow{\sim} M$. This diffeomorphism ψ in turn induces a cobordism C_ψ as in 1.2.16. This cobordism is equivalent to M.

1.3.10 Corollary. *Let $M : \Sigma_0 \Rightarrow \Sigma_1$ be a cobordism. Then there is a decomposition $M = M_{[0,\varepsilon]} M_{[\varepsilon,1]}$ such that $M_{[0,\varepsilon]}$ is diffeomorphic to a cylinder over Σ_0. (And similarly there is another decomposition such that the part near Σ_1 is diffeomorphic to a cylinder over Σ_1.)*

Indeed, take a Morse function $f : M \to [0, 1]$, and let t be the first critical value. Then for $\varepsilon < t$ the interval $[0, \varepsilon]$ is regular, so if we cut M along the inverse image $f^{-1}(\varepsilon)$, by the regular interval theorem we get the required decomposition

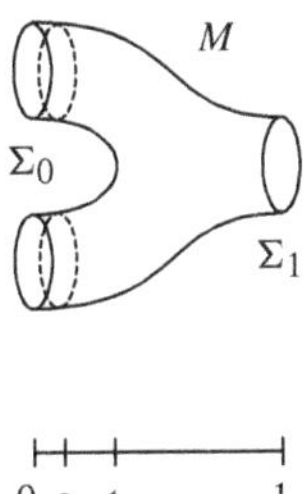

1.3.11 Gluing of general cobordisms. Given cobordisms $M_0 : \Sigma_0 \Rightarrow \Sigma_1$ and $M_1 : \Sigma_1 \Rightarrow \Sigma_2$, take Morse functions $f_0 : M_0 \to [0, 1]$ and $f_1 : M_1 \to [1, 2]$, and consider the topological manifold $M_0 \coprod_{\Sigma_1} M_1$, with the induced continuous map $M_0 \coprod_{\Sigma_1} M_1 \to [0, 2]$. Choose $\varepsilon > 0$ so small that the two intervals $[1 - \varepsilon, 1]$ and $[1, 1 + \varepsilon]$ are regular for f_0 and f_1 respectively; then the inverse images of these two intervals are diffeomorphic to cylinders. So within the interval $[1 - \varepsilon, 1 + \varepsilon]$ we are in the situation of 1.3.7, and we can take the smooth structure to be the one coming from the cylinder. The result is a gluing $M_0 M_1$ of the two cobordisms $M_0 : \Sigma_0 \Rightarrow \Sigma_1$ and $M_1 : \Sigma_1 \Rightarrow \Sigma_2$.

We have shown there is always a smooth manifold $M_0 M_1$ which is homeomorphic to $M_0 \coprod_{\Sigma_1} M_1$ and whose smooth structure agrees with each part. But the construction was not canonical. It is not clear a priori that we could not make other choices and end up with a smooth structure on $M_0 \coprod_{\Sigma_1} M_1$, not isomorphic to the first one.

The arguments given in 1.3.5 can be formalised to prove the following general result. (See Milnor [36], Theorem 1.4, and compare also Theorem 8.2.1 in Hirsch [27].)

1.3.12 Theorem. *Let Σ be an out-boundary of M_0 and an in-boundary of M_1, and consider the topological manifold $M_0 M_1 := M_0 \coprod_{\Sigma} M_1$. Let α and β be two smooth structures on $M_0 M_1$ which both induce the original structure on M_0 and M_1 (via pullback along the inclusion maps). Then there is a diffeomorphism $\phi : (M_0 M_1, \alpha) \xrightarrow{\sim} (M_0 M_1, \beta)$ such that $\phi \mid_{\Sigma} = \mathrm{id}_{\Sigma}$.*

In other words, the smooth structure on $M_0 M_1$ is unique up to diffeomorphism. (More details on these questions can be found in Munkres [40], Chapter 6.)

1.3.13 Digression: homeomorphism versus diffeomorphism. These constructions relate to two very profound questions in differentiable topology. The first question is: given a topological manifold, does it admit a smooth structure? Second: if it admits a smooth structure, is it then unique up to diffeomorphism? The second question was answered negatively by Milnor [35] (1957) who found an example of a smooth 7-manifold (called an exotic sphere) which is homeomorphic to the usual 7-sphere but not diffeomorphic to it. Work of Freedman (1982) and Donaldson (same year) on 4-manifolds implies that there are also exotic $\mathbb{R}^4$s, and that there exist 4-manifolds which do not admit any smooth structure (see Lawson [31]).

In these notes we are concerned mostly with 2-dimensional cobordisms – in that case the above subtleties vanish: *every topological surface admits a smooth structure, and two smooth surfaces are diffeomorphic if and only if they are homeomorphic*. (This result goes back to the 1920s (Rado, Kerékjartó, ...), see Moise [37].)

1.3.14 The composition of two cobordism classes. So far we have shown that given specific cobordisms $M_0 : \Sigma_0 \Rightarrow \Sigma_1$ and $M_1 : \Sigma_1 \Rightarrow \Sigma_2$ then there is a well defined diffeomorphism class $M_0M_1 : \Sigma_0 \Rightarrow \Sigma_2$.

We must now check that the result does not depend on the actual cobordisms chosen, but only on their class, so that it makes sense to compose cobordism *classes*. But in a indirect way we have already shown that. Suppose we have diffeomorphisms (rel the boundary) $\psi_0 : M_0 \xrightarrow{\sim} M_0'$ and $\psi_1 : M_1 \xrightarrow{\sim} M_1'$,

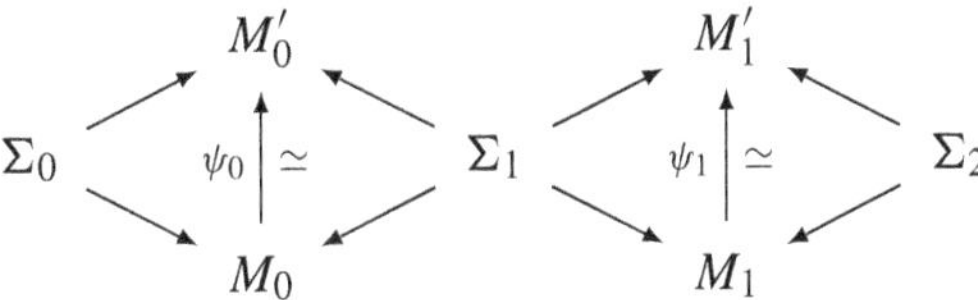

then there is a gluing M_0M_1 and a gluing $M_0'M_1'$, and also the two diffeomorphisms ψ_0 and ψ_1 glue in the category of continuous maps, so we get a homeomorphism $\psi : M_0M_1 \xrightarrow{\sim} M_0'M_1'$ which is a diffeomorphism on each piece

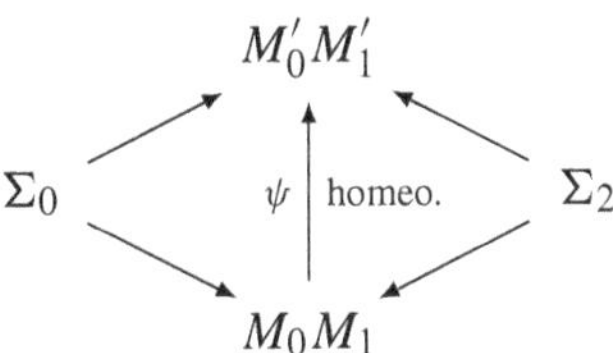

Now we can simply use this homeomorphism to *define* smooth structure on $M_0'M_1'$ – with this smooth structure, by construction, ψ is a diffeomorphism. Now this smooth structure on $M_0'M_1'$ might not be the one we started with, but according to the theorem, the two are diffeomorphic rel the boundary.

1.3.15 Associativity of the composition. We must show that given three cobordism classes, represented by

$$\Sigma_0 \overset{M_0}{\Rightarrow} \Sigma_1 \overset{M_1}{\Rightarrow} \Sigma_2 \overset{M_2}{\Rightarrow} \Sigma_3,$$

then we have the following equality of cobordism classes:

$$(M_0M_1)M_2 = M_0(M_1M_2).$$

This follows from the construction. First notice that the pushout enjoy that property:

$$(M_0 \coprod_{\Sigma_1} M_1) \coprod_{\Sigma_2} M_2 = M_0 \coprod_{\Sigma_1} (M_1 \coprod_{\Sigma_2} M_2)$$

(modulo canonical identifications which we will treat as identities). Now the smooth structure consists in replacing the coordinate charts with charts coming from cylinders, within two narrow strips near Σ_1 and Σ_2. Since these two strips are disjoint, clearly it makes no difference if first we make the replacement near Σ_1 and then the replacement near Σ_2, or if we do it the other way around.

Identity cobordisms and invertible cobordisms

We have already seen that the composition of two cylinders is again a cylinder (and more generally with cobordisms diffeomorphic to cylinders). Now we show that the class of cylinders is in fact the identity for the composition.

1.3.16 Cylinders as identity cobordisms. Intuitively, if you attach a cylinder to a boundary you do not change the topology, but in view of the subtleties sketched in 1.3.13, it is not at all obvious that the attachment does not provide new possible smooth structures. The proof is of course a variation of 1.3.12. Recall that every cobordism already decomposes into a cylinder followed by something else. Precisely, let M be our cobordism from Σ_0 to Σ_1, and let C denote a cylinder over Σ_0. We want to show that up to diffeomorphism rel the boundary, $CM = M$. Decompose M as $M = M_{[0,\varepsilon]} M_{[\varepsilon,1]}$ where the first part is diffeomorphic to a cylinder over Σ_0. Now we can finish the proof by writing (modulo diffeomorphism):

$$CM = C(M_{[0,\varepsilon]} M_{[\varepsilon,1]}) = (CM_{[0,\varepsilon]}) M_{[\varepsilon,1]} = M_{[0,\varepsilon]} M_{[\varepsilon,1]} = M.$$

Here we used the associativity of composition, and the fact that the composition of two cylinders is again a cylinder – we can find a diffeomorphism contracting $CM_{[0,\varepsilon]}$ back to $M_{[0,\varepsilon]}$.

Now we should also prove that the cylinder is an identity on the right, but this is completely analogous to what we just did.

1.3.17 Example. Let Σ be the disjoint union of two circles Σ' and Σ''. Then the identity cobordism on Σ, the cylinder $\Sigma \times I$, is the disjoint union of two cylinders (over Σ' and Σ'' respectively). Here is a picture:

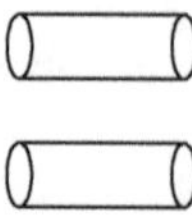

Obviously for an arbitrary Σ, the cylinder $\Sigma \times I$ has the same number of connected components as Σ.

1.3.18 Invertible cobordisms. It might seem perhaps that given a cobordism $M : \Sigma_0 \Rightarrow \Sigma_1$ we could always just take the manifold with reverse orientation $\overline{M} : \Sigma_1 \Rightarrow \Sigma_0$ to get an inverse to M. This reasoning is wrong, because that is not the definition of being an inverse. The definition is: $M' : \Sigma_1 \Rightarrow \Sigma_0$ is an inverse to $M : \Sigma_0 \Rightarrow \Sigma_1$ if MM' is the identity on Σ_0 and $M'M$ is the identity on Σ_1. It is easy to write down an example where $\overline{M}$ is not the inverse of M:

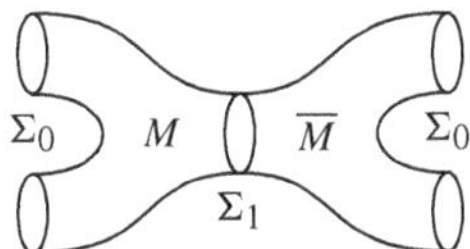

This 2-manifold is not diffeomorphic to the cylinder over Σ_0, which was drawn in Example 1.3.17 above! (The lemma below will also tell us that cannot be invertible.) For 2-dimensional cobordisms there are not so many possibilities, and in fact it is true that the only connected invertible cobordisms are the cylinders themselves, cf. 1.4.9.

It also is worth stressing that inverses must be two sided! The snake decomposition of the cylinder (1.2.21) provides an example where the composite M_0M_1 is the identity cobordism while the composite in the other order M_1M_0 is not.

1.3.19 Lemma. *Let $M : \Sigma_0 \Rightarrow \Sigma_1$ be an invertible cobordism. If M is connected (as n-manifold) then Σ_0 is connected too, and so is Σ_1.*

Proof. By assumption, the composite MM^{-1} is the cylinder $C_{\Sigma_0} = \Sigma_0 \times I$. Being a cylinder this manifold is 'horizontally connected' in the sense that every point is connected to some point on the in-boundary. But this in-boundary is also the in-boundary of M, and M is assumed to be connected, so all these points are on the same connected component. In other words, C_{Σ_0} is connected. But then by the concluding remark of 1.3.17, the base Σ_0 is also connected. (Repeating the arguments on the composite $M^{-1}M$ shows also that Σ_1 is connected.) □

1.3.20 The category *nCob*. The objects of ***nCob*** are $(n-1)$-dimensional closed oriented manifolds. Given two such objects Σ_0 and Σ_1, then an arrow from Σ_0 to Σ_1 is by definition a diffeomorphism class of oriented cobordisms $M : \Sigma_0 \Rightarrow \Sigma_1$. (In other words, the arrows are cobordism classes in the sense of 1.2.17.)

(Usually, categories are named after their objects, but this would be inconvenient here since the objects are just $(n-1)$-manifolds – we would not capture the personality of the category...)

Henceforth, the term 'cobordism' will be used to mean 'cobordism class'.

1.3.21 Some categorical digressions. So we had two problems in this section: first, composition was not well defined, and second, there was no identity, no matter how we chose the composition. Both problems were solved by passing to a quotient by an equivalence relation. In this way a lot of information was thrown away. Many people nowadays (e.g. Baez and Dolan [9]) believe that instead of throwing away this information, it would be better to invent a looser notion of category – one where these problems are part of the theory: we do not require composition of arrows to be strictly well defined, we only require it to be well defined up to some sort of equivalence. So instead of speaking about *the* composition, one could speak only of *a* composition. Such considerations lead to the notions of higher-dimensional categories where we have usual arrows (in dimension 1), arrows between arrows (dimension 2), and so on. The need for such notions comes from many different areas of mathematics, first and foremost homotopy theory. This is a very abstract subject and also very fascinating; a good place to start is John Baez's web site [8] which is a gold mine of nontechnical introductions for the nonexpert – about higher categories, and mathematical physics in general – and it is replete with precise references to the literature.

1.3.22 Diffeomorphisms and their induced cobordism classes. It was mentioned in 1.2.16 how a diffeomorphism $\phi : \Sigma_0 \xrightarrow{\sim} \Sigma_1$ induces a cobordism $C_\phi : \Sigma_0 \Rightarrow \Sigma_1$, via the cylinder construction. Take the cylinder $\Sigma_1 \times I$, and map Σ_0 onto the in-boundary via ϕ and map Σ_1 onto the out-boundary via the identity map. Alternatively, take the cylinder $\Sigma_0 \times I$, and map Σ_0 onto the in-boundary via the identity map, and map Σ_1 onto the out-boundary via ϕ^{-1}. It is easy to see that these two constructions give equivalent cobordisms:

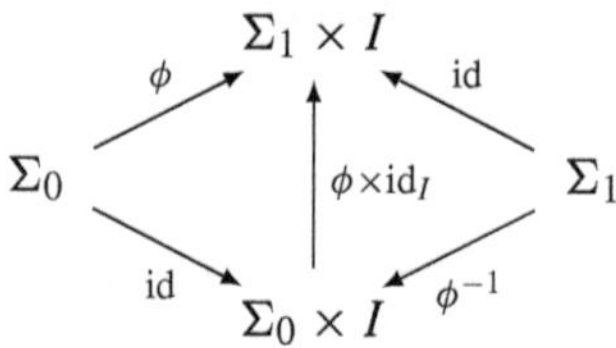

Here and elsewhere, whenever we write maps into a cylinder like this it is understood that the map on the left maps onto the in-boundary and the map on the right maps onto the out-boundary.

Using this remark, it is not difficult to prove (cf. Exercise 1 below) that if given two diffeomorphisms $\Sigma_0 \overset{\phi_0}{\tilde{\to}} \Sigma_1 \overset{\phi_1}{\tilde{\to}} \Sigma_2$, then we have

$$C_{\phi_0} C_{\phi_1} = C_{\phi_0 \phi_1}.$$

It also follows that the identity diffeomorphism $\Sigma \tilde{\to} \Sigma$ induces the identity cobordism. In other words, we have a functor from the category of $(n-1)$-manifolds and diffeomorphisms to the category ***nCob***.

In particular, a cobordism induced from a diffeomorphism is invertible.

Two questions are natural at this point. When do two diffeomorphisms give the same cobordism class? Does every invertible cobordism class arise from a diffeomorphism? The first question is settled by the next proposition. The second question we will answer (affirmatively) only in the 2-dimensional case, in the next section.

1.3.23 Proposition. *Two diffeomorphisms* $\Sigma_0 \rightrightarrows \Sigma_1$ *induce the same cobordism class* $\Sigma_0 \Rightarrow \Sigma_1$ *if and only if they are (smoothly) homotopic.*

Proof. Recall that two maps $\psi_0 : \Sigma_0 \to \Sigma_1$ and $\psi_1 : \Sigma_0 \to \Sigma_1$ are *(smoothly) homotopic* if one map can be deformed smoothly into the other, i.e. when there exists a smooth map $\Psi : \Sigma_0 \times I \to \Sigma_1$ which agrees with ψ_0 in one end of the cylinder and with ψ_1 in the other:

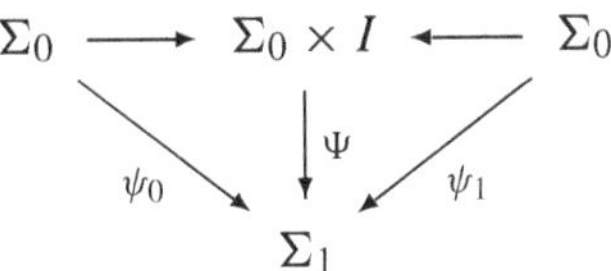

Now to have such a diagram is equivalent to having this diagram (requiring the map to be compatible with the projection to I):

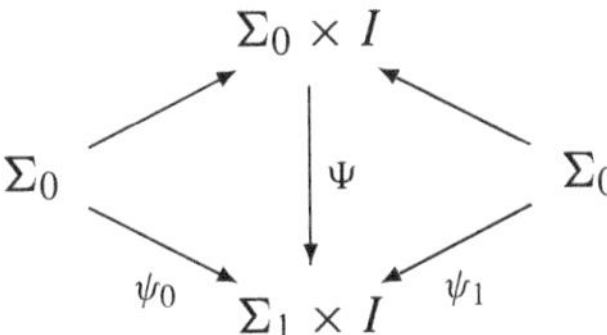

Now we claim that this diagram in turn amounts to having an equivalence of cobordisms. To see this, compose the diagram with

$$\Sigma_0 \xleftarrow{\psi_1^{-1}} \Sigma_1$$

on the right, getting

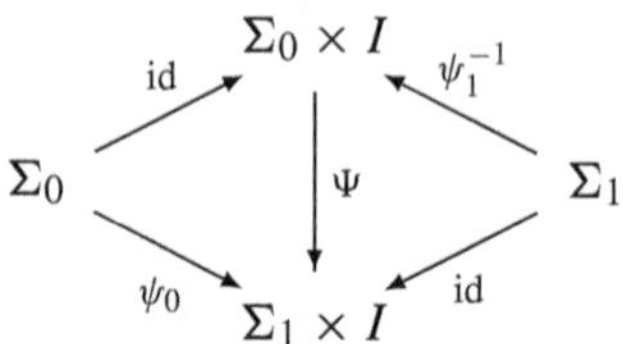

The upper part of this diagram is the cobordism class induced by ψ_1 (in the 'backward' convention); the lower part is the cobordism induced by ψ_0, and Ψ expresses that they are equivalent.

This shows that homotopic maps induce equivalent cobordisms. The other implication follows by composing $\Sigma_0 \times I \to \Sigma_1 \times I$ with the projection $\Sigma_1 \times I \to \Sigma_1$. □

So in particular, a cobordism $\Sigma \Rightarrow \Sigma$ induced by a diffeomorphism $\psi : \Sigma \xrightarrow{\sim} \Sigma$ is the identity if and only if ψ is homotopic to the identity. As an example of a diffeomorphism which is not homotopic to the identity, take the twist diffeomorphism $\Sigma \coprod \Sigma \to \Sigma \coprod \Sigma$ which interchanges the two copies of Σ.

Monoidal structure

The category structure describes how to connect cobordisms in serial, in other words, how to connect the output of one cobordism to the input of another, and so on, to make chains of cobordisms, building up larger ones from simpler ones like this

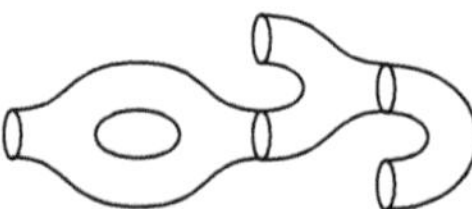

But we should be careful here: this drawing is not really a composition in the categorical sense, because the output of one cobordism does not match the input of the following! To make sense of it we need to add cylinders like this

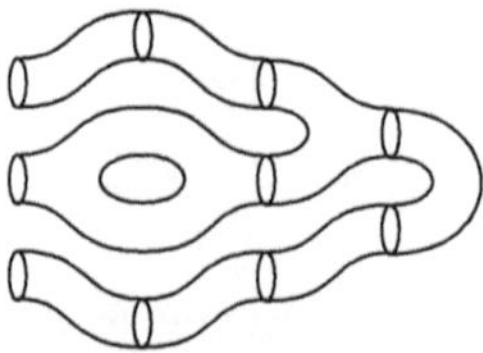

(Cylinder means: equivalent to a cylinder – they are drawn curved for graphical convenience.) Now we can truly affirm that this is the composition of these three cobordisms:

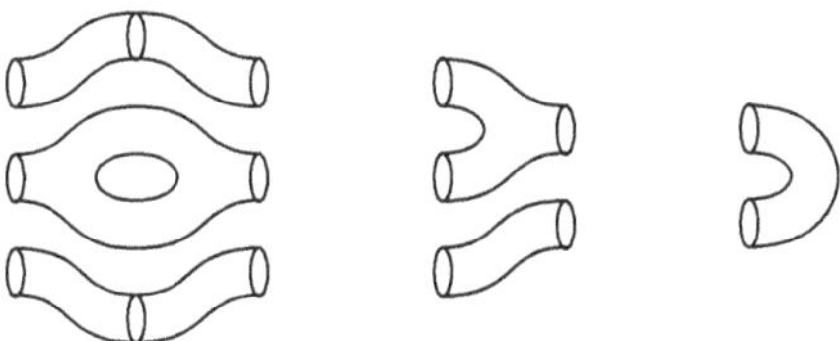

The lesson to be learned from this is that it is also important to consider parallel couplings, that is, disjoint union of cobordisms! On the next few pages we collect some basic properties of this parallel coupling, and mention that it amounts to giving *symmetric monoidal structure* to the category ***nCob***.

First there are some basic questions about disjoint union we need to settle – so let us digress into set theory. Let ***Set*** denote the category of sets and set mappings.

1.3.24 Discussion of disjoint unions of sets. Our main concern is, first, to define what exactly the disjoint union of two sets is, and second, to explain that given two sets A and B, in general it is *not* true that $A \coprod B = B \coprod A$, but that there is a canonical isomorphism between them, called the *twist map*, which is so natural that it is easily mistaken for an equality.

What exactly is the disjoint union of two sets? Let us start with a couple of remarks on the usual union of two sets which is easy to understand concretely – we will call it the *plain union*, just to avoid confusion. Given two sets $A = \{v, w, x, y\}$ and $B = \{x, y, z\}$ as in this drawing

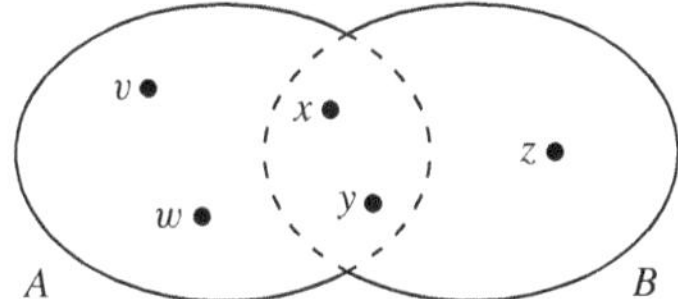

the plain union $A \cup B = \{v, w, x, y, z\}$ is obtained by removing the separating dashed lines to get one big set containing the original two sets as subsets. Clearly $A \cup B = B \cup A$. The plain union $A \cup B$ enjoys the following universal property: whenever we have inclusions $A \subset S$ and $B \subset S$ then there is induced an inclusion $A \cup B \subset S$ (unique, since set-theoretic inclusions are always unique). In fact this property characterises $A \cup B$ completely: it is the smallest set containing both A and B. In this sense, the plain union is more or less nature given – it is not a construction.

The notion of disjoint union is designed to do for general maps what the plain union does for set-theoretic inclusions – inclusions form a very restricted class of maps. In return, the notion is more abstract and somewhat more artificial.

The universal property we require is *the coproduct property* (cf. A.3.4 in the Appendix): $A\coprod B$ should be a set equipped with two structure maps $A \to A\coprod B \leftarrow B$ such that for every set S and every pair of maps $A \to S \leftarrow B$, there is a unique map $A\coprod B \to S$ such that this diagram commutes:

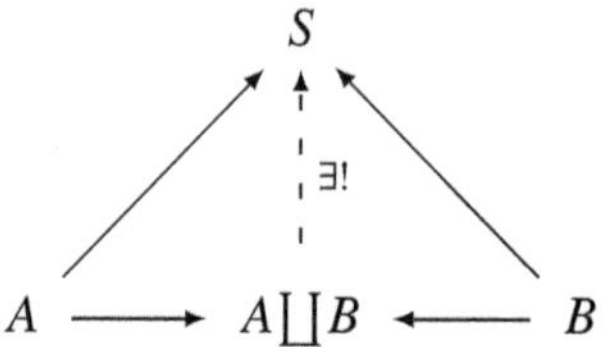

Now, A and B may have some subset in common, but nothing forces the maps to S to agree on the overlap. So in our construction of $A\coprod B$ we have to take two copies of each of the elements in the overlap and keep track of which copy belonged to which set. In other words, we need to label those twice-appearing elements – or label *all* the elements, as we shall do for the sake of uniformity. An obvious idea for doing this is to use the names of the two sets as labels, and declare the disjoint union $A\coprod B$ to be the set $\{v_A, w_A, x_A, y_A, x_B, y_B, z_B\}$. However, the names of the sets are inappropriate as labels – think of the degenerate (but important) case where we take disjoint union of two identical sets, say $A\coprod A$. What is needed is some external distinction of the two copies of A. For example we could use a positional approach, like distinguishing the two sets by calling one the left-hand set and the other the right-hand set. Otherwise we could exploit the fact that we write text in a linear way, so that one set comes before the other, and use the labels 'subscript 1' and 'subscript 2'. This has the advantage that we immediately get a notion of disjoint union of any n-tuple of sets. A side effect of this is that we introduce an *ordering* on the collection of sets involved.

Let us define $A\coprod B$ to be the set $\{v_1, w_1, x_1, y_1, x_2, y_2, z_2\}$. No matter how we choose to label, the important remark is that A and B are not subsets of this set! The point is that to consider A as a subset we would need $x = x_1$ (and $y = y_1$), and to consider B a subset we would need $x = x_2$ (and $y = y_2$), which together would imply $x_1 = x_2$ (and $y_1 = y_2$) (by transitivity of $=$), and thus we would have the plain union instead of the disjoint union!

What we *do* have in our construction are injective maps $A \hookrightarrow A\coprod B$ and $B \hookrightarrow A\coprod B$. These maps simply put the relevant index on each element, e.g. $v \mapsto v_1$. (Note that these two maps do not agree on the overlap: on A we have $x \mapsto x_1$, and on B we have $x \mapsto x_2$.) These injections are a crucial part of the structure since they provide the relation between the original sets and the new one we constructed. Using this relation it is an easy exercise to see that our construction does indeed have the universal property.

The universal property has important consequences: first of all, if we choose another labelling scheme (for example, declaring the disjoint union of A and B above to be $\{v_{\text{left}}, w_{\text{left}}, x_{\text{left}}, y_{\text{left}}, x_{\text{right}}, y_{\text{right}}, z_{\text{right}}\}$) then the universal property gives a canonical comparison bijection between these two sets. In this way it is immaterial what we choose to call the elements of the disjoint union. The important thing is that we know how to relate to the original two sets, and this information is provided by the structure injections.

Let us stick to the definition we made above, using natural numbers as labels.

We already mentioned that this labelling scheme in an obvious way generalises to a notion of disjoint union of any n-tuple of sets. We should note that this works also for $n = 0$: the disjoint union of no sets. The universal property in this case is this: the disjoint union of these 0 sets is a set P_0 equipped with maps from each of the given sets (there are none), such that for every other such set S, there is a unique map $P_0 \to S$. In other words, P_0 is the initial object in ***Set***, the empty set (cf. A.3.1).

A category with such n-ary products for all n, and satisfying certain axioms, is called a monoidal category – in Chapter 3 we will be more precise. The category $(\textbf{\textit{Set}}, \coprod, \varnothing)$ is a monoidal category.

Now we come finally to the promised remark: that $A \coprod B \neq B \coprod A$. This just means that these two sets are not identical. But what is the difference? Just that the labels 1 and 2 are interchanged. So this difference is just a special case of the freedom of choosing labels, and thus there is a bijection $\tau_{A,B}$: $A \coprod B \xrightarrow{\sim} B \coprod A$, called the twist map. The twist map does nearly nothing – if you compose with the structure injections you get the identity maps! – it only interchanges label, or, if we think of the labels as indicating position, it changes the order of the two 'factors'. Actually we get a whole family of maps: one for each pair of sets. This family is *natural*, in the specific categorical sense: it means that given two pairs of sets with arrows between them like this $f : X \to Y$ and $f' : X' \to Y'$, then it makes no difference whether we apply the twist on $X \coprod X'$ and then apply the function $f' \coprod f$ or whether we first apply the function $f \coprod f'$ and then the twist on $Y \coprod Y'$.

The twist map has some other obvious properties (e.g. $\tau_{A,B}\tau_{B,A} = \text{id}_{A \coprod B}$). With these properties it is the prototype of what is called a *symmetric structure* on a monoidal category: $(\textbf{\textit{Set}}, \coprod, \varnothing, \tau)$ is an example of a symmetric monoidal category. We will study this notion and be more precise in Chapter 3.

1.3.25 Disjoint unions of manifolds. Given two manifolds Σ and Σ', we can form their disjoint union $\Sigma \coprod \Sigma'$, which is again a manifold. If Σ and Σ' are oriented, then there is a unique orientation on $\Sigma \coprod \Sigma'$ such that the inclusion maps are orientation preserving. Just as in the category of sets, $\coprod$ is the coproduct in the category of (oriented) manifolds. Again the empty manifold is initial object. What we want to say (and we will say it with more authority in 3.2.25) is that the category of (oriented) manifolds, disjoint union and empty set is a monoidal category.

Just as for sets, there is a twist map $\tau_{\Sigma,\Sigma'} : \Sigma \coprod \Sigma' \xrightarrow{\sim} \Sigma' \coprod \Sigma$. It just interchanges the two 'factors' – it is clearly a diffeomorphism. All told, these structures turn the category of (oriented) manifolds into a symmetric monoidal category.

1.3.26 Disjoint union of cobordisms. Given two cobordisms $M : \Sigma_0 \Rightarrow \Sigma_1$ and $M' : \Sigma'_0 \Rightarrow \Sigma'_1$, then likewise we can form their disjoint union $M \coprod M'$ which is naturally a cobordism from $\Sigma_0 \coprod \Sigma'_0$ to $\Sigma_1 \coprod \Sigma'_1$.

Again we have the empty cobordism $\varnothing_n : \varnothing_{n-1} \Rightarrow \varnothing_{n-1}$.

The notion of disjoint union of two cobordism *classes* is obvious: take a representative for each cobordism and form their disjoint union. Taking other representatives clearly yields disjoint unions which are diffeomorphic (and the identity on the boundary), so the disjoint union of cobordism classes is well defined. This is to say, the triple $(\boldsymbol{n}\mathbf{Cob}, \coprod, \varnothing)$ is a monoidal category, (cf. 3.2.44).

1.3.27 Twist cobordisms and symmetric structure. We saw in 1.2.16 that every diffeomorphism $\Sigma_0 \to \Sigma_1$ induces a cobordism $\Sigma_0 \Rightarrow \Sigma_1$, via the cylinder construction. Now the cobordism corresponding to the twist diffeomorphism $\Sigma \coprod \Sigma' \xrightarrow{\sim} \Sigma' \coprod \Sigma$ will be called the twist cobordism (for Σ and Σ'), denoted $T_{\Sigma,\Sigma'} : \Sigma \coprod \Sigma' \Rightarrow \Sigma' \coprod \Sigma$. We draw it like this

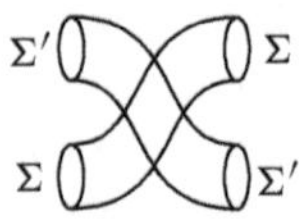

(The drawing is not meant to indicate that the two components intersect: the reason for drawing it like this instead of something like or is to avoid any idea of crossing over or under. Since we are talking about abstract manifolds, not embedded anywhere, it has no meaning to talk about crossing over or under.)

Now just as for sets and for manifolds, the twist cobordism satisfies the properties necessary to classify as a symmetric structure on the monoidal category $(\boldsymbol{n}\mathbf{Cob}, \coprod, \varnothing)$. These properties are treated in detail in Chapter 3 (page 161). The property which will be important to us here is the naturality property:

1.3.28 Lemma. *Consider two cobordisms $M : \Sigma_0 \Rightarrow \Sigma_1$ and $M' : \Sigma_0' \Rightarrow \Sigma_1'$. Then the following square commutes*

$$\begin{array}{ccc} \Sigma_0 \coprod \Sigma_0' & \overset{M \coprod M'}{\Longrightarrow} & \Sigma_1 \coprod \Sigma_1' \\ {\scriptstyle T_{\Sigma_0,\Sigma_0'}} \Downarrow & & \Downarrow {\scriptstyle T_{\Sigma_1,\Sigma_1'}} \\ \Sigma_0' \coprod \Sigma_0 & \underset{M' \coprod M}{\Longrightarrow} & \Sigma_1' \coprod \Sigma_1 \end{array}$$

Proof. Start by changing the diagram a bit by reversing the direction of the right-hand twist map – so here we depend on another crucial property of the twist map which is easy to see holds: $T_{\Sigma_1,\Sigma_1'}$ is invertible and its inverse is T_{Σ_1',Σ_1}. Now the statement amounts to comparison between the cobordism $M \coprod M'$ itself and the composite $T_{\Sigma_0 \coprod \Sigma_0'}\ (M' \coprod M)\ T_{\Sigma_1' \coprod \Sigma_1}$. Writing out the underlying smooth maps, the diagram we want to establish is this:

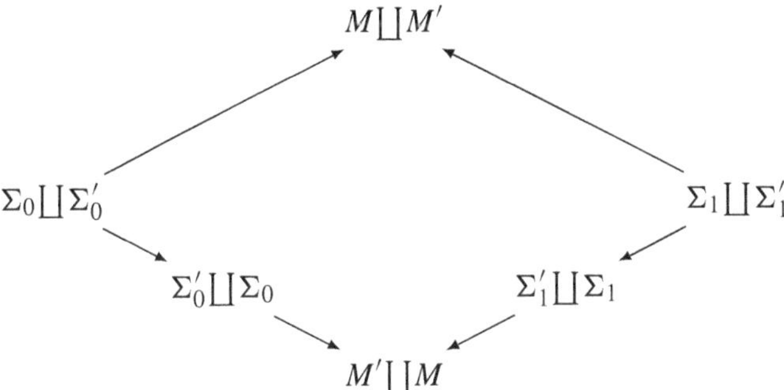

To prove that these two cobordisms are equivalent in the sense of 1.2.17,we need to exhibit a diffeomorphism from $M \coprod M'$ to $M' \coprod M$ making the diagram commute. But this is easy: the twist diffeomorphism does the job! □

As an exercise in disjoint union, let us prove this

1.3.29 Lemma. *Let $M' : \Sigma_0' \Rightarrow \Sigma_1'$ and $M'' : \Sigma_0'' \Rightarrow \Sigma_1''$ be two cobordisms (of the same dimension). If the disjoint union cobordism $M := M' \coprod M'' : \Sigma_0' \coprod \Sigma_0'' \Rightarrow \Sigma_1' \coprod \Sigma_1''$ is invertible then M' and M'' are also invertible cobordisms.*

Proof. (For simplicity we will make the extra assumption that the four sigma manifolds are nonempty.) That M is invertible means there exists a cobordism

$N : \Sigma_1' \coprod \Sigma_1'' \Rightarrow \Sigma_0' \coprod \Sigma_0''$ such that $MN = C_{\Sigma_0' \coprod \Sigma_0''} = C_{\Sigma_0'} \coprod C_{\Sigma_0''}$, the cylinder over $\Sigma_0' \coprod \Sigma_0''$. Now this cobordism is nonconnected so it follows also that N is nonconnected: it can then be written as a disjoint union $N = N' \coprod N''$ with $N' : \Sigma_1' \Rightarrow \Sigma_0'$ and $N'' : \Sigma_1'' \Rightarrow \Sigma_0''$. Now we can look at the connected components separately. For the one-primed part we see $M'N' : \Sigma_0' \Rightarrow \Sigma_1' \Rightarrow \Sigma_0'$ is the cylinder $C_{\Sigma_0'}$, and similarly the two-primed part $M''N''$ is the cylinder $C_{\Sigma_0''}$. Repeating the arguments with M and N in the converse order shows that N' is indeed an inverse to M' and that N'' is an inverse to M''. □

Using this we can prove the following generalisation of Lemma 1.3.19.

1.3.30 Lemma. *If $M : \Sigma_0 \Rightarrow \Sigma_1$ is invertible then Σ_0 and Σ_1 have the same number of connected components.*

Proof. The case where M is connected is Lemma 1.3.19. If M has more than one connected component, then by composing with a twist map (a permutation of the boundaries) if necessary, we can assume that M is the disjoint union of cobordisms with fewer connected components. Each of these is invertible by the previous lemma, and so the result follows by induction on the number of components. □

Topological quantum field theories

1.3.31 Vector spaces. Consider the category $\mathbf{Vect}_{\Bbbk}$ of vector spaces over a field $\Bbbk$ and $\Bbbk$-linear maps. Equipped with tensor product as 'paralleling', with the ground field as neutral space, and with the canonical twist map σ which interchanges the two factors of a tensor product, $(\mathbf{Vect}_{\Bbbk}, \otimes, \Bbbk, \sigma)$ is also a symmetric monoidal category. (For the definition of vector spaces and tensor product, see 2.1.1; for the definition of monoidal categories (and symmetric monoidal categories) see Chapter 3 (notably 3.2.4, 3.2.34, and 3.2.28).)

A *monoidal functor* (between two monoidal categories) is one that preserves the monoidal structure. A *symmetric* monoidal functor between two symmetric monoidal categories is one that sends the symmetry of one monoidal category to the symmetry of the other. The precise definitions are given in Chapter 3.

As already mentioned, this terminology allows for an elegant restatement of the definition of topological quantum field theory.

1.3.32 Functorial definition of topological quantum field theories. An *n-dimensional topological quantum field theory* is a symmetric monoidal functor from $(\mathbf{nCob}, \coprod, \emptyset, T)$ to $(\mathbf{Vect}_{\Bbbk}, \otimes, \Bbbk, \sigma)$.

Let us compare this definition with the first one (1.2.23). Suppose we have a TQFT in the sense of 1.2.23. Then axiom A1 implies that we have a well defined map $\mathscr{A}$ from ***nCob*** to ***Vect***$_{\Bbbk}$ (i.e. that it only depends on the class of M). Axioms A2 and A3 say that this is really a functor: it respects identities and composition.

Axioms A4 and A5 state that the functor is monoidal, namely it takes disjoint union to tensor products, and takes the neutral object $\emptyset$ to the neutral object $\Bbbk$.

Finally we should require $\mathscr{A}$ to take the symmetry to the symmetry, in order for $\mathscr{A}$ to be a *symmetric* monoidal functor. This was not stated as a requirement in the first definition of a TQFT, because it is quite difficult to imagine how it could be otherwise... but it should have been included. We will come back to this discussion in 3.3.3.

1.3.33 Historical remarks. The concept of topological quantum field theory is due to Witten [50] (1988) – see also the reference list in Atiyah's paper [5]. Mathematical axioms for TQFTs were given by Atiyah [5]. In Quinn's 1991 lectures [43], the categorical viewpoint is well developed. Quinn also takes the opportunity to generalise the whole setting: in his definition a TQFT does not only talk about cobordisms, but more generally about a *domain category for TQFT* which is a pair of categories related by certain functors and operations which play the rôle of space and space-time categories in the usual cobordism setting. In this more general setting the domain category can also be of combinatorial or algebraic nature instead of geometric or topological.

The observation 1.2.25 that the TQFT axioms actually imply finite dimensionality was probably first made by Quinn [43]. It seems that this fact has largely been overlooked in the literature on TQFTs: most texts require explicitly the vector spaces to be of finite dimension, with excuses like 'otherwise we get convergence problems with infinite sums' (cf. Example 1.2.29).

Exercises

1. Given two diffeomorphisms $\Sigma_0 \overset{\phi_0}{\underset{\sim}{\to}} \Sigma_1 \overset{\phi_1}{\underset{\sim}{\to}} \Sigma_2$ between closed $(n-1)$-manifolds, show that the cylinder construction (1.3.22) gives this equality of cobordism classes: $C_{\phi_0} C_{\phi_1} = C_{\phi_0 \phi_1}$.
2. In the situation of 1.3.26, use the universal property of $\coprod$ in the category of (oriented) manifolds to make explicit in which sense $M \coprod M'$ is naturally a cobordism from $\Sigma_0 \coprod \Sigma_1$ to $\Sigma_0' \coprod \Sigma_1'$, as claimed in 1.3.26. (Construct natural maps $\Sigma_0 \coprod \Sigma_1 \to M \coprod M'$ and $\Sigma_0' \coprod \Sigma_1' \to M \coprod M'$.)

3. In the category ***2Cob*** of all 2-dimensional cobordisms, consider a circle Σ. Consider the set $E := \mathrm{Hom}_{\boldsymbol{2Cob}}(\Sigma, \Sigma)$ of all cobordism classes from Σ to itself, and the subset H of all cobordism classes whose underlying manifold is connected. Show that E is a monoid, and that H is a submonoid.
4. (Durhuus and Jónsson [20].) Show that the following notion of direct sum of TQFTs is again a TQFT. The *direct sum* $\mathscr{A}$ of two n-dimensional TQFTs $\mathscr{A}'$ and $\mathscr{A}''$ associates to each connected $(n-1)$-manifold Σ the direct sum vector space: $\Sigma\mathscr{A} := \Sigma\mathscr{A}' \oplus \Sigma\mathscr{A}''$, and to each nonconnected $(n-1)$-manifold it associates the tensor product of the vector spaces associated to its connected components. To each connected n-cobordism $M : \Sigma_0 \Rightarrow \Sigma_1$ it associates the direct sum linear map $M\mathscr{A} := M\mathscr{A}' \oplus M\mathscr{A}'' : \Sigma_0\mathscr{A} \to \Sigma_1\mathscr{A}$, and to each nonconnected M, it associates the tensor product of the maps associated to its connected components.

1.4 Generators and relations for *2Cob*

The categories ***nCob*** are very difficult to describe for $n \geq 3$. But the category ***2Cob*** can be described explicitly, and that is the goal of this section. (The reason for this difference is of course that while there is a complete classification theorem for surfaces, no such result is known in higher dimensions.)

So ***2Cob*** is the category whose objects are the closed oriented 1-manifolds, and whose arrows are the diffeomorphism classes of oriented cobordisms between them. The aim of this section is to describe a set of generators and relations for this category. What does this mean?

Preliminary observations

We will come to categories in a short moment, but first, to warm up, let us consider an example of generators and relations for a *group*, the symmetric group. We will need this result anyway.

1.4.1 Generators and relations of a group. Let G be a finite group. A generating set for G is a subset $S \subset G$ such that every element in G can be written as a product of elements in S (and their inverses). A relation (or a rewriting rule) is the equality of two ways of writing a given element in terms of the generators. A set of relations R is complete if every other relation that holds in G can be established by combining the relations of R.

1.4.2 Moore's theorem. (Cf. Moore [38] (1897). See Coxeter and Moser [13] for a more accessible reference.) Let $\mathfrak{S}_k$ denote the symmetric group on $k \geq 4$ letters $\{x_1, \ldots, x_k\}$. Then $\mathfrak{S}_k$ is generated by the transpositions that interchange two adjacent letters,

$$\tau_i := (x_i \, x_{i+1}) \qquad i = 1, \ldots, k-1,$$

subject to the following relations:

$$\begin{aligned} \tau_i \tau_i &= \mathrm{id} \\ \tau_i \tau_j \tau_i &= \tau_j \tau_i \tau_j \qquad \text{for } j = i+1 \\ \tau_i \tau_j &= \tau_j \tau_i \qquad \text{for } j > i+1. \end{aligned}$$

(It is also possible to present $\mathfrak{S}_k$ by only two generators, e.g. $(x_1 \, x_2)$ and $(x_1 \, x_2 \, \cdots \, x_k)$, but then the relations are more complicated (cf. [13], 6.2), and that presentation is not well suited for our purposes.)

Now we can think of the symmetric group $\mathfrak{S}_k$ as the category of invertible maps from the set $\{x_1, \ldots, x_k\}$ to itself. If we draw the set $\{x_1, \ldots, x_k\}$ as a column of dots, then we get the following sort of pictures for a bijective set mapping (i.e. element in $\mathfrak{S}_k$). Here we draw the three generators for $\mathfrak{S}_4$:

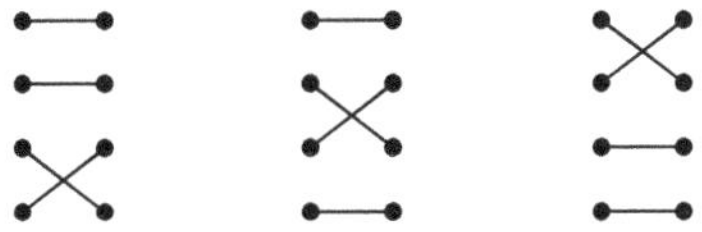

And here are all the relations:

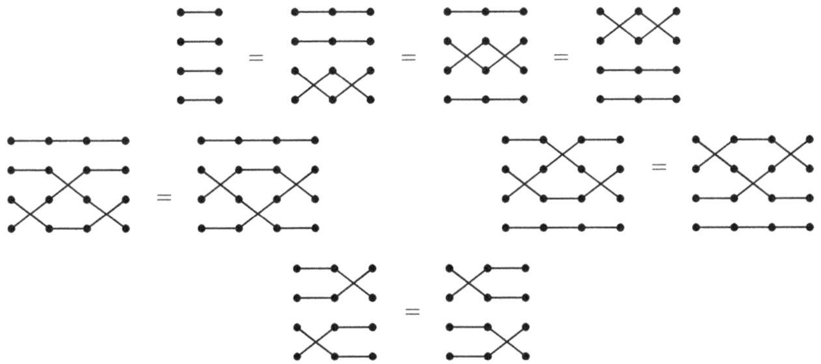

=

=

=

=

=

=

1.4.3 Paralleling as generating concept. We notice that the generators in turn are just variations of this transposition (element in $\mathfrak{S}_2$):

combined in parallel with identity permutations. If we allow paralleling as a generating concept, then there is only need for one single generator, namely that one. Similarly, using parallel coupling, all the relations can be obtained from these two:

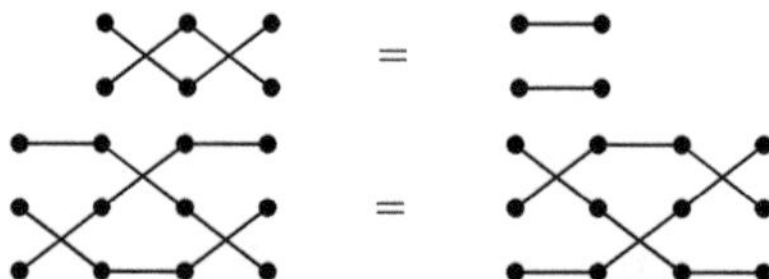

(Arranging things in parallel is really like taking their disjoint union. In this way, what we are seeing here is actually part of a monoidal structure. However, if we take two elements of $\mathfrak{S}_2$ and couple them in parallel, what we get is not a new element in $\mathfrak{S}_2$ but rather an element in $\mathfrak{S}_4$. So in order to really get something monoidal we should take not only $\mathfrak{S}_k$, the category of invertible maps from $\{x_1, \ldots, x_k\}$ to itself, but rather the category of all the invertible maps on any of the sets $\{x_1, \ldots, x_k\}$, with k running from 0 to infinity... Compare 3.4.23, and don't miss Exercise 2 at the end of this section.)

1.4.4 From groups to categories. We mentioned that the symmetric group $\mathfrak{S}_k$ can be interpreted as the category of all invertible maps from a certain set $S = \{x_1, \ldots, x_k\}$ to itself. In other words, it is a subcategory of ***Set***, which comprises exactly one object (namely S) and all of whose arrows are invertible. This observation exploited the idea that the elements in the group $\mathfrak{S}_k$ are already maps of a sort. But in fact (cf. 3.1.11 and 3.1.19), any group can be interpreted as a category (a category with only one object, and such that all arrows are invertible – the group elements correspond to the arrows in the category). Via this observation we are led to the concept of

1.4.5 Generators and relations for a category. A *generating set* for a category $\boldsymbol{C}$ is a set S of arrows such that every arrow in $\boldsymbol{C}$ can be obtained by composing the arrows of S. A relation is the equality of two ways of writing a given arrow in terms of the generators. A set R of relations is complete if every relation can be obtained by combining the relations in R.

For large categories, like the category of vector spaces or the category of cobordisms, there are too many objects to get hold of a generating set. (In fact, the word 'large' has a precise technical meaning: these categories have so many objects that they do not even form a set. A generating 'set' of arrows will always have at least as many elements as there are objects, so it will not be a set either. While these questions are interesting enough we will not go into them in these notes...)

Before looking for generators of a large category like this, it is necessary to cut it down to a more manageable size – this is common sense, independent of the more philosophical considerations of the parenthesis above. The way to do that is the standard construction of taking a skeleton of a category.

1.4.6 Skeletons of a category. A *skeleton* of a category $\boldsymbol{C}$ is a full subcategory comprising exactly one object from each isomorphism class.

A skeleton $\boldsymbol{Z} \subset \boldsymbol{C}$ captures the essential structure of $\boldsymbol{C}$ in the sense that it is equivalent to $\boldsymbol{C}$ (cf. A.2.8): the embedding $\boldsymbol{Z} \hookrightarrow \boldsymbol{C}$ is full, faithful, and essentially surjective by construction. The reason why it is called a skeleton is that it is a minimal category with this property: if we took a subcategory of a skeleton with fewer objects we would lose essential surjectivity, and if instead we took fewer arrows we would lose fullness. The skeleton construction is not canonical however: in general, there is no canonical way of choosing a representative from each isomorphism class, but the various possible skeletons are always isomorphic. (An isomorphism between two skeletons can be constructed by means of the isomorphisms between the representatives.)

1.4.7 Example from linear algebra. While all this may sound very abstract, it is actually something that we do every day without thinking so much about it. As a brief example, consider the category of real vector spaces (of finite dimension) and linear maps. We know that every vector space of dimension n is isomorphic to $\mathbb{R}^n$, so as skeleton we take the subcategory consisting of the vector spaces $\mathbb{R}^n$, $n \geq 0$. The linear maps between these spaces are given by the m-by-n matrices. What are the generators now? Well, we know that every matrix (of size m-by-n and of rank r, say) can be written as a matrix product (=composition of linear maps) of form ADB where A and B are invertible square matrices (of size m and n respectively) and D is a matrix consisting of zeros, except for an r-by-r minor identity matrix. The invertible matrices in turn can be written as a product of elementary matrices. So a set of generators for the category of vector spaces $\mathbb{R}^n$ are the elementary matrices together with those nonsquare matrices of form D as described...

So instead of finding generators for the large category ***2Cob*** we will content ourselves with finding generators for a skeleton.

1.4.8 The objects of *2Cob*. The first observation is that every closed oriented 1-manifold is diffeomorphic to a finite disjoint union of circles – remember that closed implies compact. More concretely, we can fix one specific circle Σ and affirm that every connected closed 1-manifold is diffeomorphic to Σ, and that every closed 1-manifold with n connected components is diffeomorphic to the disjoint union of n copies of Σ.

Note that even if we have a given circle and a copy of it with reverse orientation, there is an orientation-preserving diffeomorphism between them: to be concrete, take the two copies and place them in the plane at each side of a line, and at equal distance; now reflection in the line provides an orientation-preserving diffeomorphism between them:

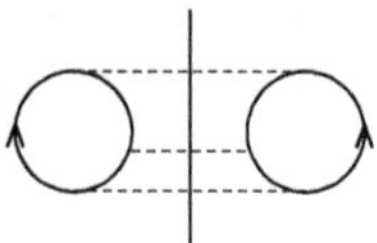

The second observation is trickier:

1.4.9 Proposition. *Two closed oriented* 1*-manifolds* Σ_0 *and* Σ_1 *are diffeomorphic if and only if there is an invertible cobordism between them.*

Proof. We have seen that given a diffeomorphism we can use the cylinder construction 1.2.16 to get an invertible cobordism. For the converse statement, notice first that by the first observation, it is enough to show that if there is an invertible cobordism $M : \Sigma_0 \Rightarrow \Sigma_1$ then Σ_0 and Σ_1 have the same number of connected components. But this was proved in 1.3.30. □

In conclusion, two objects of ***2Cob*** (i.e. two closed 1-manifolds) are in the same isomorphism class of ***2Cob*** (i.e. there exists an invertible cobordism between them) if and only if they have the same number of connected components. In fact more is true: *the only invertible* 2*-cobordisms are those induced from diffeomorphisms.*

Recall that two diffeomorphisms induce the same cobordism class if and only if they are homotopic. Now in dimension 1, it is easy to classify all diffeomorphisms up to homotopy: every orientation-preserving diffeomorphism from the circle to itself is homotopic to the identity. (This can be seen for example by noting that the winding number classifies the homotopy classes of maps $S^1 \to S^1$, and only winding number 1 can be an (orientation-preserving) diffeomorphism.) So in general, for a closed 1-manifold, up to homotopy the only diffeomorphisms are the permutations of its connected components. Hence the only invertible 2-cobordisms are the permutation cobordisms (with the identity permutation corresponding to the identity cobordism).

(Anticipating the themes of the next section, here is another argument why the only invertible cobordism from a circle to a circle is the cylinder: the *genus* of a composite is at least the sum of the genera, so only cobordisms of genus 0 can be invertible...)

1.4.10 Skeleton of *2Cob*. So we get a skeleton of ***2Cob*** as follows. Let **0** denote the empty 1-manifold; let **1** denote a given circle Σ, and let **n** denote the disjoint union of n copies of Σ. Then the full subcategory $\{\mathbf{0}, \mathbf{1}, \mathbf{2}, \dots\}$ is a skeleton of ***2Cob***. (The arrows are all the possible cobordisms between these objects.)

By abuse of notation we will denote this category ***2Cob***, or otherwise we make the abusive convention that when we speak of generators and relations for ***2Cob*** we really mean generators and relations of this skeleton.

1.4.11 Generators for a monoidal category. Notice that the chosen skeleton is closed under the operation of taking disjoint union, and that in fact disjoint union is the main principle of its construction. Since now we have this monoidal structure we should use it, so instead of allowing only composition as the engine for generating new cobordisms from old ones, we will also allow disjoint union. This means we are really talking about an ampler notion of generating set.

A *generating set for a monoidal category* $\boldsymbol{C}$ is a set S of arrows such that every arrow in $\boldsymbol{C}$ can be obtained from the arrows in S by combining composition and 'monoidal paralleling'. (This last term is nothing official: in our case it means disjoint union.) In figurative terms, a generating set is a set of simple building blocks from which every cobordism can be obtained by parallel and serial connection.

For example, the cylinder over a disjoint union of two circles is itself the disjoint union of two cylinders (over a circle) so if is in the set of generators we do not need .

1.4.12 The twist. Since we have included disjoint union as one of the allowed operations by which we generate, we can largely concentrate on cobordisms which are connected. But not completely: for each object **n**, ($n \geq 2$), there are cobordisms $\mathbf{n} \Rightarrow \mathbf{n}$ which are not the identity. For example, for **2** (the disjoint union of two circles), we have the *twist* (cf. 1.3.27):

It is important to notice that the twist is *not* the disjoint union of two identity cobordisms: precisely, it is not equivalent to the disjoint union of two cylinders, since the diffeomorphisms realising an equivalence are required to respect the boundary. The point is that the two copies of Σ can be distinguished, as explained in 1.3.24 (otherwise there would be no meaning in saying they are disjoint!) Usually we make this distinction by calling one copy 'the first copy' and the other 'the second copy'. But we could also choose a point in $\Sigma \coprod \Sigma$ and distinguish the two copies by telling which one contains the point and which one does not. In this viewpoint, the true identity (the cylinder over $\Sigma \coprod \Sigma$) has the property that this point stays on the same connected component of the cobordism, while in the twist it changes component, so therefore the two cannot be diffeomorphic.

Generators

Let us state right away the results about generators.

1.4.13 Proposition. *The monoidal category* **2Cob** *is generated under composition (serial connection) and disjoint union (parallel connection) by the following six cobordisms:*

It should be noted that usually one does not include identity arrows when listing generators for a category, because identity arrows are automatically in every category, by definition. This is similar to the usage for groups or monoids: one says $(\mathbb{N}, +, 0)$ is generated by 1, without listing 0: the element 0 can be written as a sum of 1s, namely the empty sum, so in this sense 0 is even generated by the empty set! Here we include as generator just because it makes it easier to think of cutting up surfaces in pieces. Since we include this superfluous generator, we will also get some corresponding extra relations, cf. 1.4.24.

We give two proofs of Proposition 1.4.13, since they both provide some insight. In any case some nontrivial result about surfaces is needed. The first proof relies directly on the classification of surfaces (quoted below): the connected surfaces are classified by some topological invariants, and we simply build a surface with given invariants! To get the nonconnected cobordisms we use disjoint union and permutation of the factors of the disjoint union. Since every permutation can be written as a composition of transpositions, the sixth generator suffices to do this. The drawback of this first proof is that it does not say so much about how a given surface relates to this 'normal form' – this information is hidden in the quoted classification theorem.

The second proof relies on a result from Morse theory (which is an ingredient in one of the possible proofs of the classification theorem (cf. Hirsch [27])), and here we do exactly what we missed in the first proof: start with a concrete surface and cut it up in pieces; now identify each piece as one of the generators. (Philosophy: If you have something complicated it is easier to split it apart than to assemble it from its parts. . .)

1.4.14 Genus and Euler characteristic of surfaces. Let us first give a couple of reminders on surfaces. The *genus* of a compact, connected, oriented surface is intuitively the number of holes. So a sphere has genus 0, and a torus genus 1. For a surface with boundary, the genus is defined to be the genus of the closed surface obtained by sewing in discs along each boundary component. So a disc also has genus 0, because if you sew in a disc along its boundary you get a sphere. With this definition it is obvious that the genus does not detect anything related to the boundary. To this end the *Euler characteristic* is better.

One way to define the Euler characteristic is in terms of triangulations: if V, E, F are the numbers of vertices, edges and faces of a triangulation of a surface M then the Euler characteristic is

$$\chi(M) = V - E + F.$$

So the Euler characteristic of a disc is 1, and the Euler characteristic of a sphere is 2. Notice that removing a disc from a surface amounts to decrementing F, so the Euler characteristic drops: the cylinder has Euler characteristic 0; the pair-of-pants has -1. The relation between the Euler characteristic and the genus g of a surface is

$$\chi(M) + k = 2 - 2g$$

where k is the number of missing discs (i.e. the number of boundary components).

The Euler characteristic enjoys a cutting property: if $M = A \cup B$ then $\chi(M) = \chi(A) + \chi(B) - \chi(A \cap B)$. This is particularly convenient if M is a 2-dimensional cobordism. Then decomposing it $M = M_0 M_1$ amounts to cutting along circles S^1, and we have $\chi(S^1) = 0$ (as you can easily compute via a triangulation), so the outcome is this formula:

$$\chi(M) = \chi(M_0) + \chi(M_1).$$

Finally, we should mention that the Euler characteristic of a surface M can be computed by taking a Morse function $f : M \to I$ and summing over all the critical points. The formula is

$$\chi(M) = \sum_{x \text{ critical}} (-1)^{\text{index}_f(x)}.$$

1.4.15 Topological classification of surfaces. The classical result classifying surfaces without boundary is (cf. Hirsch [27], Theorem 9.3.5.):

Two connected, compact oriented surfaces without boundary are diffeomorphic if and only if they have the same genus (or equivalently: the same Euler characteristic).

(Note that the reflection argument of 1.4.8 shows, also for surfaces, that there is always an orientation-reversing diffeomorphism from any surface to itself, so reversing the orientation does not provide anything new.)

If a compact surface has boundary then this boundary is the disjoint union of finitely many circles, each bounding a 'missing disc'. The following result shows that these missing discs can move around freely without changing the topology:

Two connected, compact oriented surfaces with oriented boundary are diffeomorphic if and only if they have the same genus and the same number of in-boundaries and the same number of out-boundaries.

The Euler characteristic can detect boundary components, but not their orientation, so to classify surfaces via the Euler characteristic we still need to specify how many of the boundaries are in and how many are out.

1.4.16 'Normal form' of a connected surface. It is convenient to introduce the *normal form* of a connected surface with m in-boundaries, n out-boundaries, and genus g. It is actually a decomposition of the surface into a number of basic cobordisms. The normal form has three parts: the first part (called the in-part) is a cobordism $\mathbf{m} \Rightarrow \mathbf{1}$; the middle part (referred to as the topological part) is a cobordism $\mathbf{1} \Rightarrow \mathbf{1}$; and the third part (the out-part) goes $\mathbf{1} \Rightarrow \mathbf{n}$.

Before giving the precise description, let us draw a figure of the normal form in the case $m = 5$, $g = 4$, and $n = 4$.

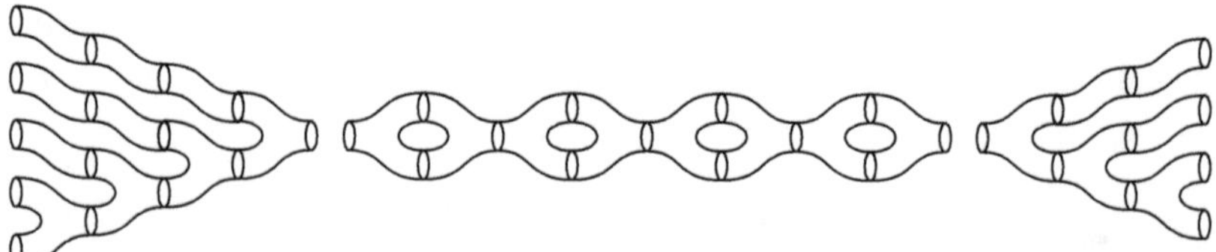

Let us describe the in-part. Suppose first that $m > 0$. Take $m - 1$ copies of and compose them, together with the appropriate number of cylinders, in such a way that the output of one connects to the lower input hole of the following. In other words, the involved cylinders always come on top of the pair-of-pants, in the disjoint union sense. In the case $m = 0$, instead of having any pair-of-pants, the whole in-part just consists of a single.

The out-part is described similarly: for $n > 0$ we have the composition of $n - 1$ copies of 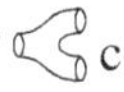 composed in such a way that the lower output hole of each piece connects to the input hole of its sequel. (So again, the cylinders needed to complete the picture appear on top of the.) In the case $n = 0$, the out-part consists of a single copy of.

Finally, the topological part consists of g pieces of type; this gives a piece with one in-boundary, one out-boundary, and genus g. (Here you can either think of g as the number of holes, or you can measure it in a more formal way by computing the Euler characteristic – exploiting the fact that $\chi(M) = \chi(M_0) + \chi(M_1)$, as we will do in 1.4.36.)

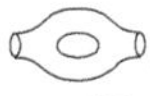

The normal form is at the same time a recipe for constructing any *connected* cobordism from the generators. Thus we have:

1.4.17 Lemma. *Every connected 2-cobordism can be obtained by composition and disjoint union of the generators* , , ,  , . □

1.4.18 Nonconnected cobordisms. If a cobordism M is not a connected manifold it is the disjoint union of connected manifolds. But this is not enough to prove Proposition 1.4.13, because that result refers to the specific notion of disjoint union of cobordisms, which implies compatibility with disjoint union of the boundaries. The easiest example of this distinction is the twist: we explained in 1.4.12 that although the twist (as a manifold) is the disjoint union of two cylinders, it is *not* the disjoint union (as cobordism) of two identity cobordisms. But by permuting the boundaries we can fix that, as we now explain in more detail.

Let $M : \mathbf{m} \Rightarrow \mathbf{n}$ be a cobordism. So it is a manifold whose in-boundary is $\Sigma_1 \coprod \Sigma_2 \coprod \cdots \coprod \Sigma_m$ and whose out-boundary is $\Sigma'_1 \coprod \cdots \coprod \Sigma'_n$. All the sigmas are just copies of one and the same circle Σ, but we have given them different names just to record their position in the disjoint unions – according to our discussion in 1.3.24.

For simplicity, assume M has two connected components, M_0 and M_1. The in-boundary on M_0 is a subset $\mathbf{p}$ of $\Sigma_1 \coprod \Sigma_2 \coprod \cdots \coprod \Sigma_m$, and the in-boundary of M_1 is the complement $\mathbf{q}$ of this subset, but there is no reason why these subsets should consist of the p first circles and the q last ones. But we can just permute the components of $\mathbf{m}$: take a diffeomorphism $\mathbf{m} \xrightarrow{\sim} \mathbf{m}$ which places the subset $\mathbf{p}$ before the subset $\mathbf{q}$. This diffeomorphism induces a cobordism S, and we can now look at the cobordism $SM : \mathbf{m} \Rightarrow \mathbf{n}$ instead. Let $(SM)_0$ and $(SM)_1$ denote the two connected components of this manifold. Now SM has the property that its in-boundary is the disjoint union of the in-boundaries of $(SM)_0$ and $(SM)_1$.

Applying the same arguments to the out-boundary of M (= the out-boundary of SM) we can also find a permutation cobordism $T : \mathbf{n} \Rightarrow \mathbf{n}$ such that altogether $SMT : \mathbf{m} \Rightarrow \mathbf{n}$ is a cobordism which is the disjoint union of its connected components – as a cobordism.

Consider this example:

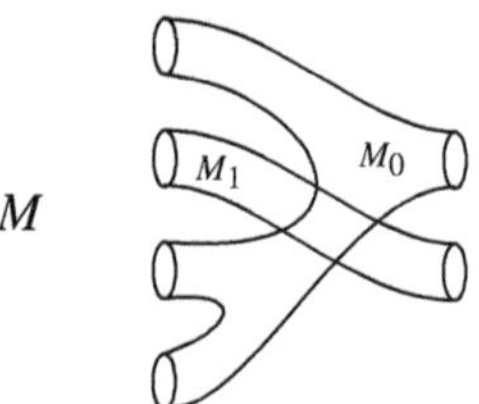

M is not the disjoint union cobordism of its connected components. But we can permute the boundaries by composing with two cobordisms S and T:

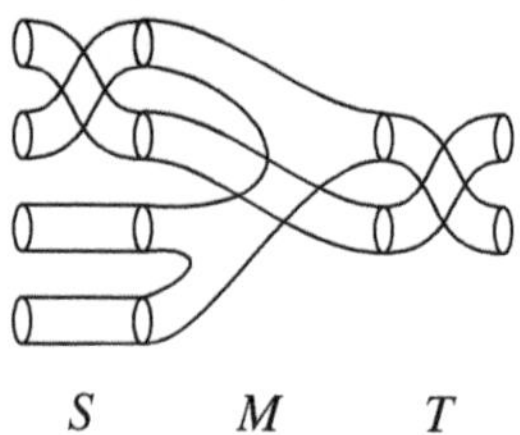

and this cobordism SMT *is* the disjoint union of its connected components $(SMT)_0$ and $(SMT)_1$:

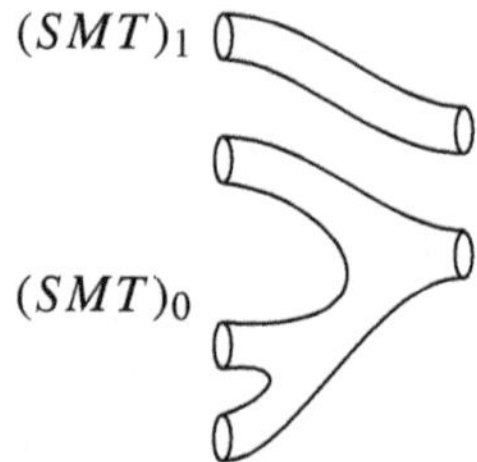

We have shown:

1.4.19 Lemma. *Every 2-cobordism factors as a permutation cobordism, followed by a disjoint union of connected cobordisms, followed by a permutation cobordism.* □

(These permutation cobordisms are the inverses of those used in the arguments above. . .)

Combining these results we get a

1.4.20 First proof of Proposition 1.4.13. By Lemma 1.4.19, every 2-cobordism factors into permutation cobordisms and a disjoint union of connected cobordisms. By Lemma 1.4.17, the connected pieces can be written in terms of the listed generators. Finally, since the symmetric groups are generated by transpositions (cf. 1.4.2 and 1.4.3), the permutation cobordisms can be obtained by composition and disjoint union of the twist cobordism (together with cylinders). □

To be specific with our example above:

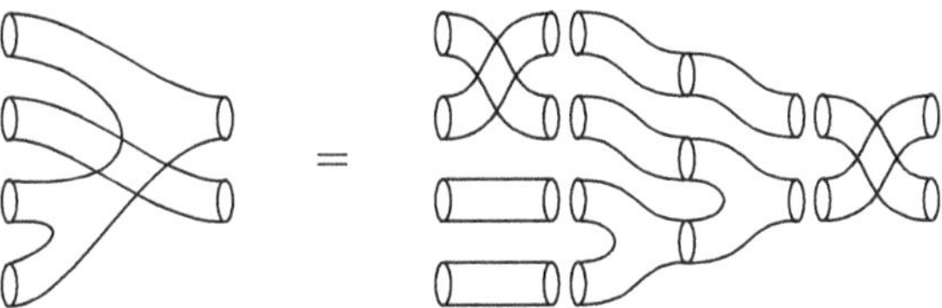

(Of course one can often find more direct decompositions – for example,

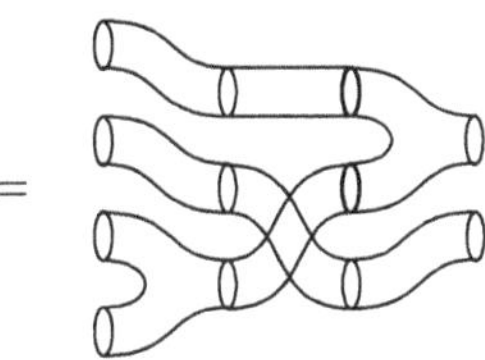

is another way of writing the example above.)

The Morse theoretic proof of Proposition 1.4.13 is different in spirit. The key is to characterise the generators in terms of their critical points. Part of this task was done in the regular interval theorem (1.3.8), where we saw that a cobordism admitting a Morse function without critical points is diffeomorphic to a cylinder. Now this was a diffeomorphism rel only one of the boundaries, so it is not enough to conclude that the cobordism is equivalent to the identity cobordism. But in any case we noted that such a cobordism $\Sigma_0 \Rightarrow \Sigma_1$ is induced by a diffeomorphism $\psi : \Sigma_0 \xrightarrow{\sim} \Sigma_1$, and 2-cobordisms induced by diffeomorphisms are equivalent to permutation cobordisms. So we have the following:

1.4.21 Corollary to the regular interval theorem (1.3.8). *If a cobordism admits a Morse function without critical points then it is equivalent to a* *permutation cobordism.* And thus it can be built from the twist cobordism (and the identity).

The next ingredient is this lemma, which we state without proof.

1.4.22 Lemma. *(See Hirsch [27], 4.4.2.) Let M be a compact connected orientable surface with a Morse function $M \to [0, 1]$. If there is a unique critical point x, and x has index 1 (i.e. is a saddle point) then M is diffeomorphic to a disc with two discs missing (these three boundaries are over 0 and 1). In other words we have*

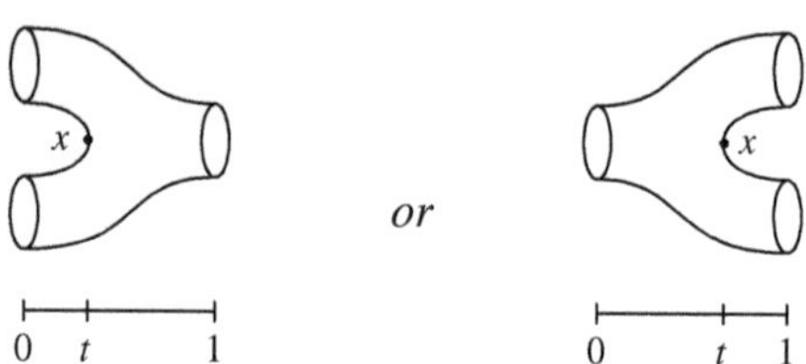

1.4.23 Morse theoretic proof of Proposition 1.4.13. Consider a cobordism $M : \Sigma_0 \Rightarrow \Sigma_1$, and take a Morse function $f : M \to [0, 1]$ with $f^{-1}(0) = \Sigma_0$ and $f^{-1}(1) = \Sigma_1$. Take a sequence of regular values $a_0, a_1, \ldots, a_k$ in such a way that there is (at most) one critical value in each interval $[a_i, a_{i+1}]$; consider one of these intervals, $[a, b]$. We can assume there is at most one critical point x in the inverse image $M_{[a,b]}$. The piece $M_{[a,b]}$ may consist of several connected components: (at most) one of them contains x; the others are equivalent to permutation cobordisms, according to Corollary 1.4.21. These pieces can be chopped up further into twist cobordisms and identities (cf. once again the observation that transpositions generate the symmetric groups).

So we can assume $M_{[a,b]}$ is connected and has a unique critical point x. Now if x has index 0 then we have a local minimum, and then $M_{[a,b]}$ is a disc like this:

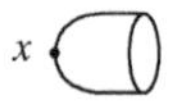

If the index is 2 then we have a local maximum, and $M_{[a,b]}$ is a disc like this:

x

And finally if the index is 1 then we have a saddle point, and by Lemma 1.4.22, $M_{[a,b]}$ is then topologically a disc with two holes, so in our picture it looks like one of these:

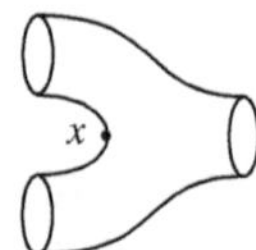

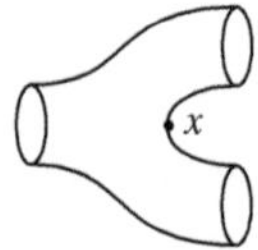

□

Relations

We will first list all the relations. Afterwards we prove that they hold, and provide some more comments on each relation.

1.4.24 Identity relations. First of all, we have already shown that the cylinders are identities. This gives a bunch of relations:

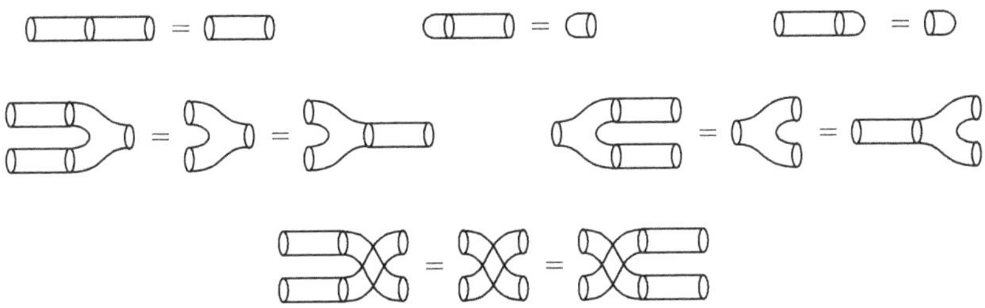

(Note that is the identity for **2**.)

1.4.25 Sewing in discs. *The following relations hold.*

(Basically, we are just sewing a disc in one of the holes of the pair-of-pants, but while this operation is fairly easy to grasp, it is not a composition of cobordisms! Indeed, is a cobordism from **0** to **1**, while goes from **2** to **1**, so it makes no sense to compose them. For this reason we need to put in a cylinder as well.)

1.4.26 'Associativity' and 'coassociativity'. *These relations hold:*

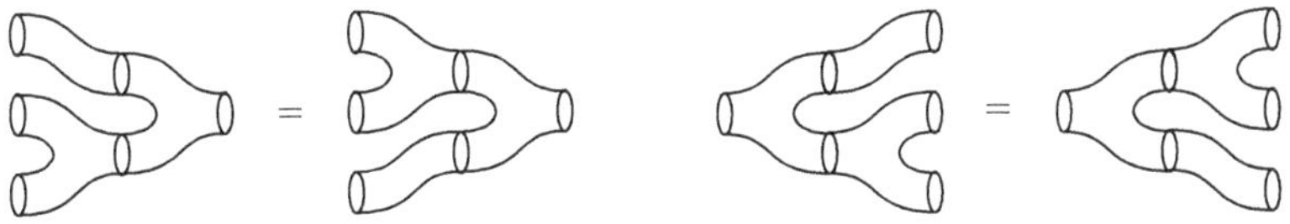

1.4.27 'Commutativity' and 'cocommutativity'. *We have:*

And finally:

1.4.28 'The Frobenius relation' *holds:*

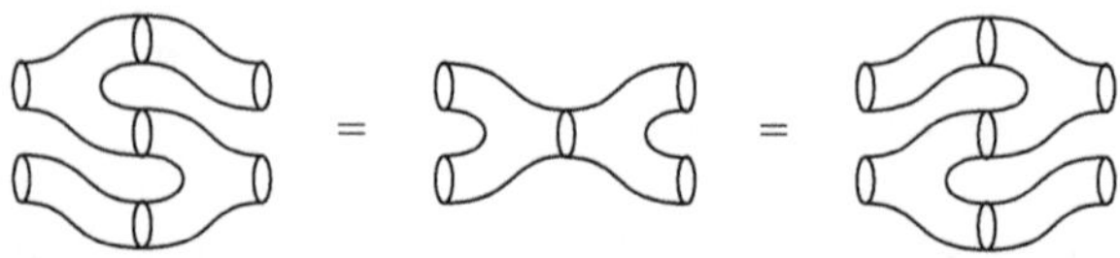

1.4.29 Easy proof of all the above relations. Simply note that in each case the surfaces have the same topological type, so according to the classification theorem they are diffeomorphic: they all have genus 0 and they have the same number of in-boundaries and out-boundaries. □

Here is another viewpoint which is perhaps more enlightening.

1.4.30 The viewpoint of nested discs. Let us here be concerned with the left-hand versions of the relations 1.4.24–1.4.26, those involving and . They are the surfaces having a single out-boundary. The cap has no other boundaries, so it is a disc; the cylinder also has an in-boundary, so it is a disc with a missing disc; and the pair-of-pants is a disc with a further two discs missing. With the orientation of their boundaries they look like this:

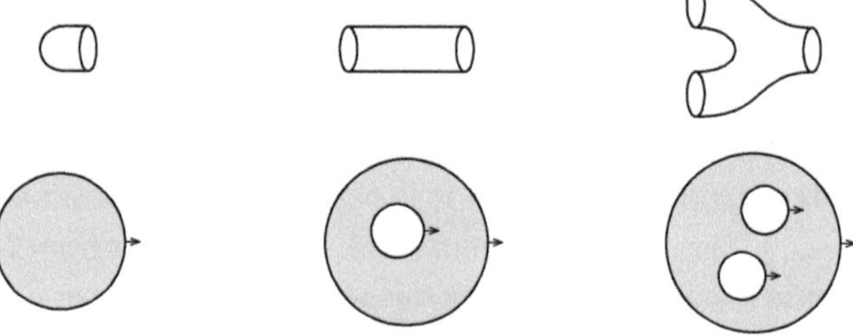

With this graphical representation it is easier to realise the relations via decompositions. In the following three paragraphs we only prove the left-hand version of the relations. Reversing the orientation gives the right-hand relations.

1.4.31 Proof of relation 1.4.25 via nested discs. This relation comes about by decomposing the cylinder, cutting it along the disjoint union of two circles, like this:

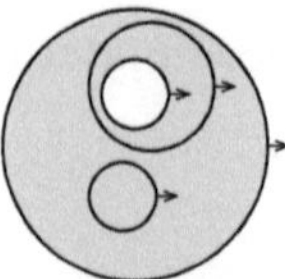

This we interpret as the composition 'sewing in a cylinder and a disc in a pair-of-pants':

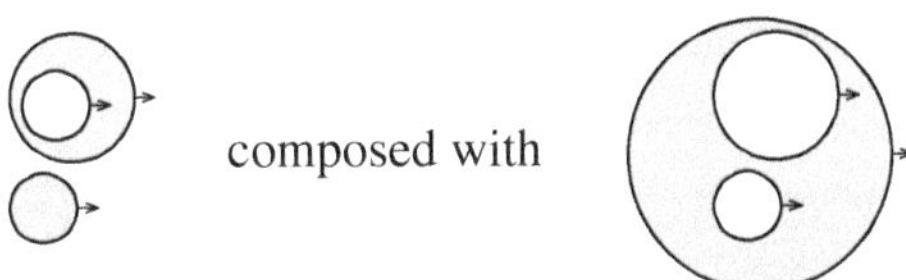

1.4.32 Proof of 1.4.26. Similarly, the associativity relation 1.4.26 is obtained by making two different decompositions of the disc with three missing discs:

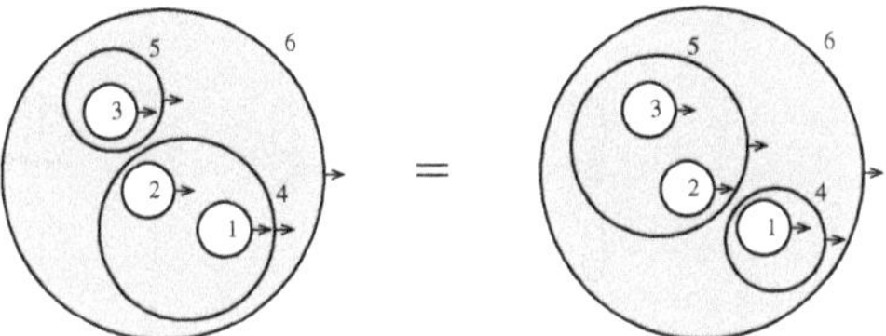

The small numbers are just to facilitate comparison with our usual associativity equation:

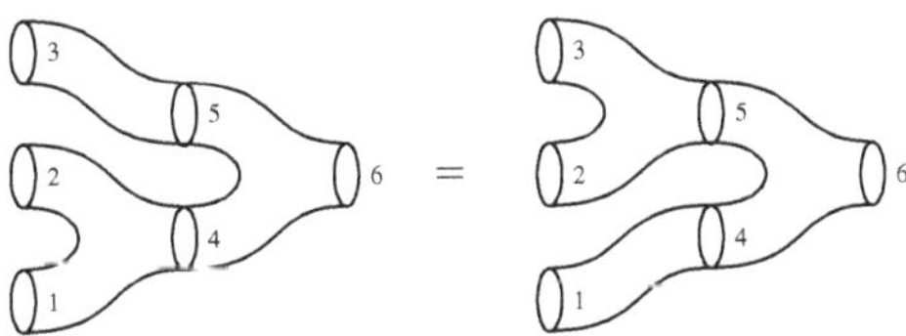

1.4.33 Commutativity in terms of nested discs. The commutativity relation 1.4.27 is difficult to draw, but easy to understand: it amounts to the fact that we can move the two in-boundaries around freely in

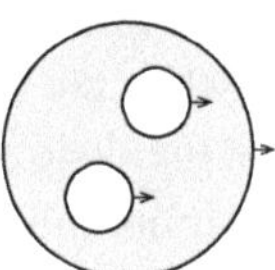

Hereby we have exhausted the possibilities for composing with itself and with , and we have also exhausted the possibilities for composing with itself and with . Now we must see whether there are any relations for combinations of these, in particular combinations of and .

1.4.34 Decomposition proof of 1.4.28. To start with, let us notice that the associativity relation 1.4.26 can be drawn like this:

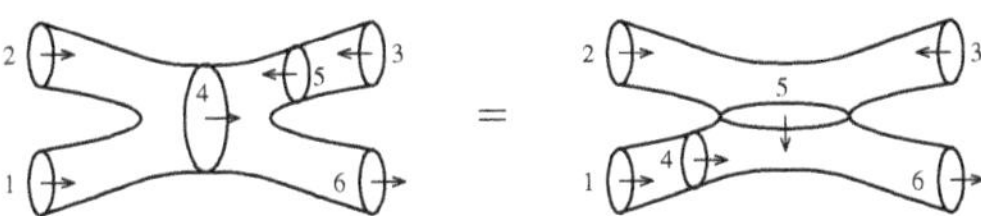

where the small numbers serve for comparison with the pictures in 1.4.30.

Reversing the orientation of the boundary with number 3 (and changing its number to 7), we get a surface which can be cut in these three ways:

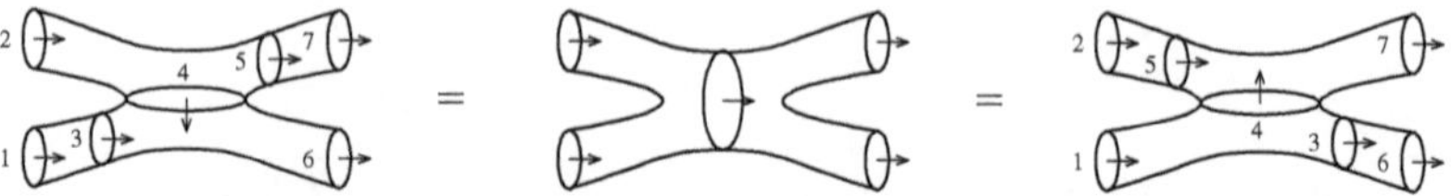

to give a decomposition proof of 1.4.28, which is repeated here with numbering, for convenience:

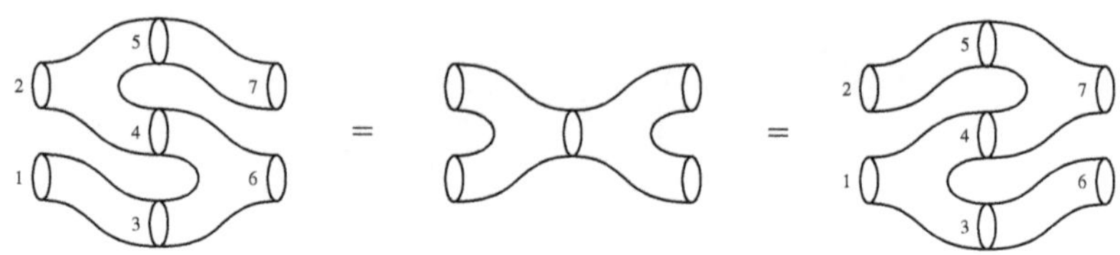

Relations involving the twist

The statement that the twist cobordism makes (***2Cob***, $\coprod$, $\emptyset$) into a symmetric monoidal category amounts to a set of relations involving the twist.

The basic relation is the fact that the twist is its own inverse

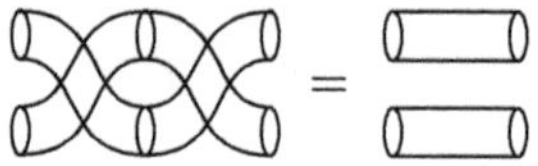

1.4.35 Relations expressing the naturality of the twist. The naturality of the twist cobordism states that for any pair of cobordisms, it makes no difference whether we apply the twist before their disjoint union or after. It is enough to describe the relations that arise when the two cobordisms are taken among the generators. Furthermore, since the disjoint union of two cobordisms can be realised as a composition of disjoint union with identity cobordisms, it is enough to state the relations for the case of a generator in disjoint union with an identity. So the following relations express the naturality, cf. 1.3.28.

First the relations of moving a twist past a cap:

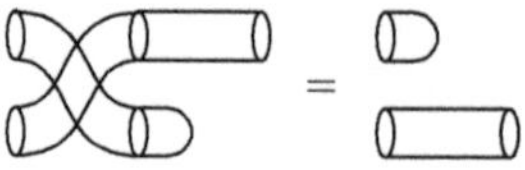

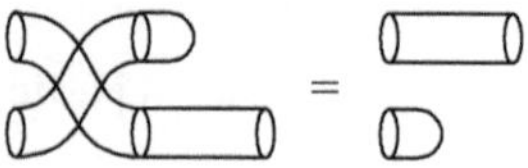

Note that these two relations are dependent in the sense that one can be derived from the other modulo the basic twist relation = .

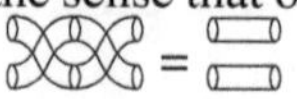

And here are the corresponding relations, with the other cap:

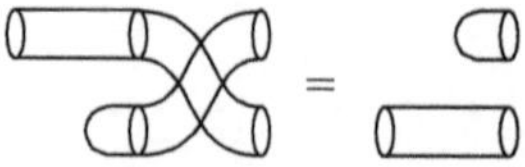

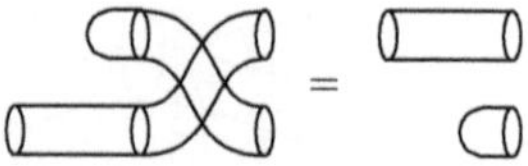

Now move the twist past the multiplication pair-of-pants:

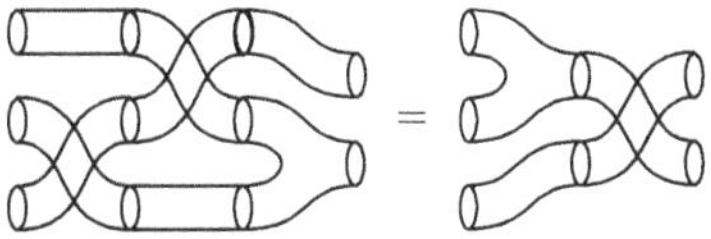

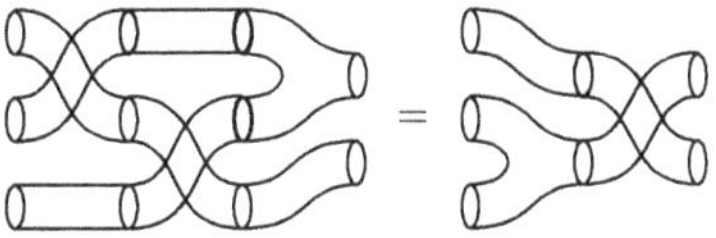

(Again these two are in fact dependent modulo the basic twist relation.)

Of course we can equally well move the twist past the comultiplication pair-of-pants:

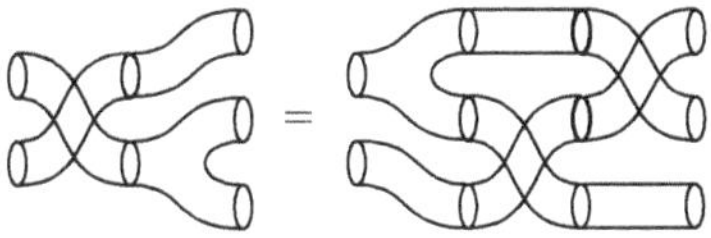

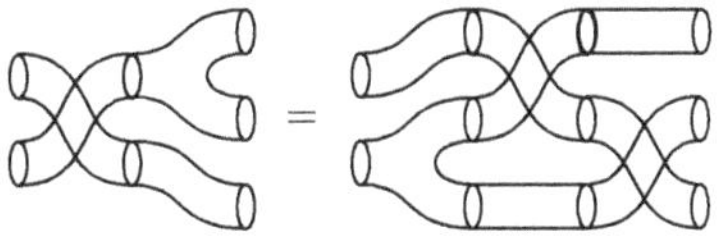

(Which are dependent modulo 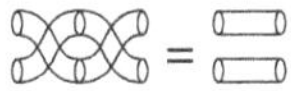 .)

Finally there is the 'symmetric group relation', which expresses how the twist moves past a twist:

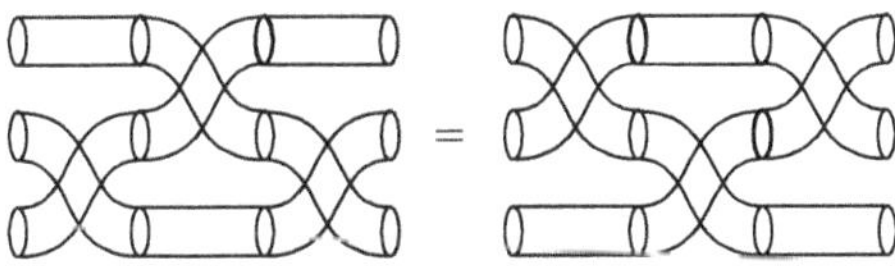

(These moves are analogous to the famous Reidemeister moves in knot theory (see Kassel [29], page 248), but they are much simpler because in our context there is no distinction between passing over or under.)

Sufficiency of the relations

Now that we have established a lot of relations, it is natural to ask whether we have enough, or whether there are other relations that we have not found yet. Of course there are infinitely many relations we could write down, so the question is whether those we have are sufficient to relate every possible decomposition of a given cobordism.

In general, it can be difficult to show such completeness results. Often the technique is to introduce some sort of normal form for the expressions as we did in 1.4.16, and then check that the listed relations are sufficient to transform any general expression into normal form.

The normal form is for connected surfaces, so we start with the case of connected surfaces. Also, we see that there are no twist cobordisms in the normal form, so this actually tells us that the twist is not needed as generator if one only considers connected surfaces. In the arguments below, we start by assuming there are no twists, and after we have settled this no-twists case we treat the general case by induction on the number of twists – in other words we eliminate the twists one by one.

1.4.36 Counting the pieces. Let us start with an arbitrary decomposition of a connected surface M with m in-boundaries, of genus g, and with n out-boundaries. The Euler characteristic is $\chi(M) = 2 - 2g - m - n$.

Let a be the number of pieces in the decomposition; let b be the number of ; let p be the number of ; and let q be the number of . By the additive property of the Euler characteristic we can now also write $\chi(M) = p + q - a - b$. Thus we have the equation

$$2 - 2g - m - n = p + q - a - b.$$

On the other hand we have the distinction between in- and out-boundaries. Summing up what each piece contributes to the number of boundaries we get the equation

$$a + q + n = b + p + m.$$

Combining the two equations we can solve for a and b to get

$$a = m + g - 1 + p$$
$$b = n + g - 1 + q,$$

and all the involved symbols are non-negative integers.

1.4.37 Moving pieces left. Our strategy is to take $m - 1$ pieces and move them to the left, until they come before any , then we will have formed the in-part of the normal form. So what do we meet on our way left? If we meet a then by the unit relation 1.4.25 we get a cylinder which we can ignore. This happens p times, so we have enough copies of to spend with that. (Note that since the surface is connected, in fact every occurrence of must be to the left of a ...)

We can also meet a . This can happen in two ways:

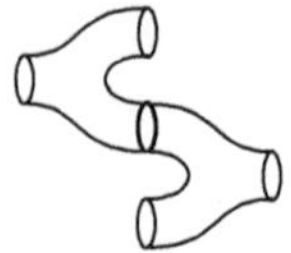

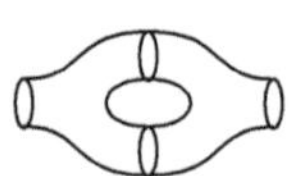

In the first case we can use relation 1.4.28 to move to the left of . The second case is trickier – we have no relations that pronounce themselves on this situation, so we will have to leave it at that. We have produced a handle; this will happen g times because each handle represents a 'genus hole'.

So among the initial $m + g - 1 + p$ copies of moving left, p of them will meet a and vanish; g of them will get stuck against a , and the remaining $m - 1$ will pass through all the way to form the in-part of the normal form. Well, we still have to argue that a can pass through one of those inert handles: use first associativity and then the Frobenius relation.

Now do the same thing moving right, until the out-part is in normal form.

We are left with something in the middle which has one in-boundary and one out-boundary. So in this part there must be the same number of and . And the left-most must be a (or perhaps a cylinder – we do not care about cylinders). So take a and move it left until it stops against this , forming a handle. Eventually we arrive at a chain of handles which is exactly the normal form.

1.4.38 Example. To get the idea of how to eliminate twist maps, this example is in a sense all we need to know:

Note carefully how each move is one of the relations. The first one is cocommutativity 1.4.27, the second move is naturality of the twist map (moving past a) cf. 1.4.35; next comes the Frobenius relation 1.4.28, and finally we use cocommutativity again.

1.4.39 Eliminating twist maps. We continue considering only connected surfaces. Let there now be a decomposition also involving twist maps. Pick one twist map T in the decomposition

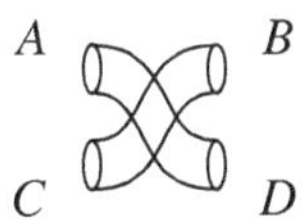

Here A, B, C, D denote the rest of the surface. Parallel with T there are other pieces, but we can always insert suitable identity maps according to the relations listed in 1.4.24, in order to ensure that all the pieces (possibly zero) parallel with T are just cylinders.

Since the surface is assumed to be connected, some of the regions A, B, C, D must be connected with each other. Suppose A and C are connected to each other. Then together they form a connected surface involving strictly less twist than the original, so by induction we can assume it can be brought on normal form using the relations. In particular only the out-part of the normal form of A-with-C touches T, and we can shuffle the pieces up and down until there is a piece which matches exactly with T like this

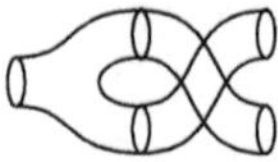

and then we use the cocommutativity relation 1.4.27 to remove T and we are done.

If B and D are interconnected the same argument applies.

So we can assume that A and B are connected to each other. Now each region A and B comprises fewer twist maps than the whole, so we can assume they are on normal form. Now in particular, the situation close to T is this:

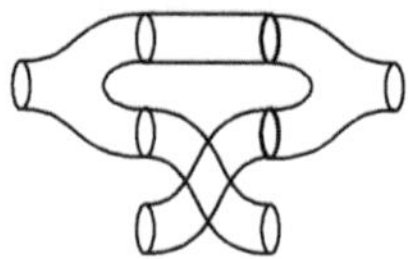

This case was treated in Example 1.4.38, so we are done.

1.4.40 Nonconnected surfaces. To prove sufficiency of the relations for nonconnected surfaces we should first define a normal form for such surfaces. It could be something factorised in three parts, for example: permutation cobordism – disjoint union of connected surfaces on normal form – and permutation cobordism again. It is easy to see that the normal form of each connected component is well defined, so modulo the order of these components this gives a sort of normal form for the middle piece. The problem of ordering the connected components, as well as the question of organising the two permutation parts concern only twist cobordisms, and we have already noted that the listed relations are precisely the relations for the symmetric groups, and we know these are sufficient.

So now the argument runs like this: starting with any 2-cobordism M built up of the six generators, we know from the arguments preceding 1.4.19 that there is a pair of permutation cobordisms S and T, such that SMT is a disjoint union of its connected components. Now the four cobordisms S^{-1}, S, T, T^{-1} can each be built up from twist cobordisms, and the fact that $S^{-1}S = \text{id}$ and $TT^{-1} = \text{id}$ can be established using the relations of the symmetric group. This means that inserting $S^{-1}S$ and TT^{-1} to get $M = S^{-1}SMTT^{-1}$ can be achieved by using the two relations

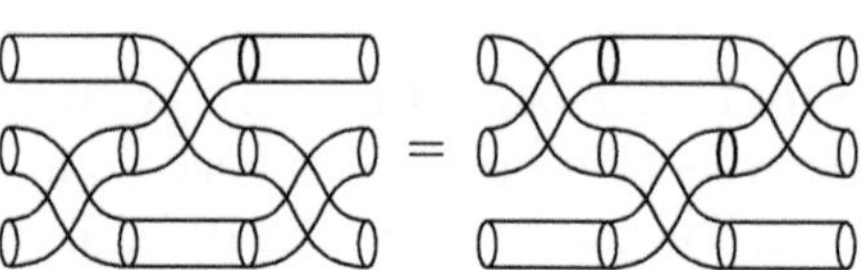

Now each of the connected components in the middle piece SMT can be brought on normal form by the arguments of 1.4.36–1.4.39, whence M has been brought in 'normal form'. As observed, this normal form is not unique, but any two such normal forms differ only by permutations, and we know that the two twist map relations above suffice to realise any permutation.

We should mention that, in fact the set of relations we have found is not minimal (cf. Exercise 6 below – it will also follow from 2.3.24):

1.4.41 Proposition. *The Frobenius relation 1.4.28 together with the unit and counit relations 1.4.25 imply the associativity and coassociativity relations 1.4.26.* □

Exercises

1. Write down generators and relations for the alternating group on four letters, in the style of 1.4.2, and also as in 1.4.3.
2. Let $\mathfrak{S}_k$ denote the category of all bijective maps from the set $\{x_1, \ldots, x_k\}$ to itself. Let Σ denote the disjoint union of all the categories $\mathfrak{S}_k$, $k \in \mathbb{N}$. Show that Σ is a skeleton for the category ***FinSet***$_0$ of all finite sets and bijective maps.

⋆3. This is a continuation of Exercise 3 on page 56. There we defined H to be the monoid of all connected cobordism (classes) from the circle **1** to itself. Use the surface classification theorem to show that this monoid is isomorphic to $(\mathbb{N}, +)$. (Hint: use the 'handle operator'  .)

4. Prove the statements made in 1.4.35, that each of the right-hand equations can be obtained from the corresponding left-hand equation together with the basic twist relation.
5. Just from the topology it is clear that this relation holds:

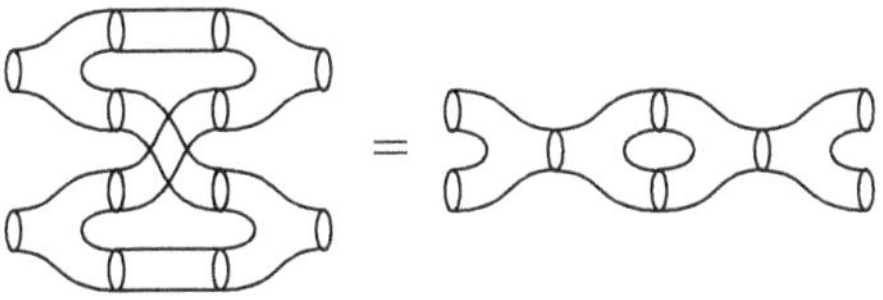

   =

   Show how to obtain this relation from the relations listed in 1.4.24–1.4.28 and 1.4.35, in the style of Example 1.4.38.
6. Prove Proposition 1.4.41, that the Frobenius relations 1.4.28 together with the 'unit'/'counit' relations 1.4.25 imply the 'associativity' relation 1.4.26 (and the 'coassociativity' too).

2
Frobenius algebras

Summary

A first preliminary section reviews some basic notions of vector spaces, pairings, algebras and modules, and establishes notation and terminology.

Section 2.2 is devoted to 'classical' theory of Frobenius algebras. A Frobenius algebra can be characterised equivalently as: a finite-dimensional algebra A equipped with an associative nondegenerate pairing, or equipped with a linear functional whose nullspace contains no nontrivial ideals, or equipped with an A-linear isomorphism to the dual space A^*. Then we give a long list of examples of Frobenius algebras. Some of these examples require more algebra than presumed elsewhere in the text, but don't panic! – these examples are not really needed elsewhere in the text.

The main result of this chapter is established in Section 2.3. It is yet another equivalent characterisation of Frobenius algebras: a Frobenius algebra is an algebra which is also a coalgebra, with a compatibility between multiplication and comultiplication. This compatibility condition is actually of topological nature, and a second important goal of this chapter is to develop a graphical language for the algebraic operations involved, which provides important insight in the structures.

In Section 2.4 we collect some results on the category of Frobenius algebras: we observe that Frobenius algebra homomorphisms are always invertible, and that the tensor product of two Frobenius algebras is again a Frobenius algebra in a canonical way. Finally we make a digression on Hopf algebras and compare their axioms with those for Frobenius algebras.

So a Frobenius algebra is a vector space with a certain structure. In Chapter 3 we are going to distil the structure and do away with the space – only the pure essence will be left! In order to prepare for this, we will from the outset describe some of the structures in a slightly more categorical manner than would normally be considered necessary.

Frobenius algebras were first studied by Frobenius [22] around 1900. The material of Section 2.2 goes back at least to Nakayama [41] in the 1930s. The characterisation of Frobenius algebras in terms of comultiplication goes back at least to Lawvere [32] (1967), and it was rediscovered by Quinn [43] and Abrams [1] in the 1990s.

2.1 Algebraic preliminaries

We assume the reader is familiar with vector spaces and tensor products, and also with rings and modules. The brief review which follows is mostly included in order to establish terminology and notation, and to introduce some slightly unusual viewpoints which are convenient for our purposes.

Vector spaces, duals, and pairings

2.1.1 Vector spaces and linear maps. Throughout this chapter we fix a ground field $\Bbbk$. (The reader can safely think of $\Bbbk$ as being $\mathbb{R}$ or $\mathbb{C}$ or whatever she prefers.)

A *vector space* over $\Bbbk$ is an abelian group V (written additively) equipped with a $\Bbbk$-action

$$V \times \Bbbk \to V.$$

This map is required to satisfy these axioms:

$$\begin{aligned} (v + v')c &= vc + v'c & & v, v' \in V,\ c \in \Bbbk \\ v(c + c') &= vc + vc' & & v \in V,\ c, c' \in \Bbbk \\ v(cc') &= (vc)c' & & v \in V,\ c, c' \in \Bbbk \\ v1 &= v & & v \in V. \end{aligned}$$

Note in particular that $\Bbbk$ itself is a vector space. The action is simply multiplication in $\Bbbk$.

A *linear map* between two vector spaces V and W is a group homomorphism $\phi : V \to W$ which commutes with the $\Bbbk$-action:

$$\begin{array}{ccc} V \times \Bbbk & \xrightarrow{\phi \times \mathrm{id}_\Bbbk} & W \times \Bbbk \\ \downarrow & & \downarrow \\ V & \xrightarrow[\phi]{} & W \end{array}$$

Let $\textbf{\textit{Vect}}_\Bbbk$ denote the category of vector spaces and linear maps.

2.1.2 Bilinear maps and tensor products. A *bilinear map* is a map $V \times W \to T$ such that for each fixed $v \in V$ the resulting map $W \to T$ is linear, and for each fixed $w \in W$, the resulting map $V \to T$ is linear.

Given two vector spaces V, W, their tensor product $V \otimes W$ is a vector space which is universal for bilinear maps in the following sense: it comes equipped with a bilinear map $V \times W \to V \otimes W$ such that every bilinear map $V \times W \to T$ factors uniquely $V \times W \to V \otimes W \to T$, where $V \otimes W \to T$ is linear. For an explicit construction of the tensor product, see Lang [30].

Similarly there are notions of multilinear maps, and the universal objects are the tensor products $V_1 \otimes \ldots \otimes V_k$. We could also build this up iteratively from the tensor product in two variables: we always identify any ways of setting parentheses

$$(V_1 \otimes \ldots \otimes V_k) \otimes (W_1 \otimes \ldots \otimes W_l) \simeq V_1 \otimes \ldots \otimes V_k \otimes W_1 \otimes \ldots \otimes W_l.$$

There is also a special tensor product with zero factors. By definition this is the universal linear function in no variables. This is $\Bbbk$ itself, so by the rules above we get canonical identifications

$$\Bbbk \otimes V \simeq V \simeq V \otimes \Bbbk.$$

2.1.3 Conventions. In this chapter, all tensor products are over $\Bbbk$, and when we write $\operatorname{Hom}(V, W)$ we always mean $\operatorname{Hom}_{\Bbbk}(V, W)$, the space of $\Bbbk$-linear maps from V to W.

2.1.4 Linear functionals. A linear map from a vector space V to the ground field $\Bbbk$ is called a *linear functional* (or a linear form).

2.1.5 Duals. The space of linear functionals on V is called the *dual* of V, and is denoted

$$V^* := \operatorname{Hom}(V, \Bbbk).$$

Given a linear map $\psi : V \to W$, the *dual map* is

$$\begin{aligned} \psi^* : W^* &\to V^* \\ \Lambda &\mapsto \psi\,\Lambda \end{aligned}$$

where $\psi\,\Lambda$ denotes the composite $V \xrightarrow{\psi} W \xrightarrow{\Lambda} \Bbbk$.

In this way, taking the dual on vector spaces and linear maps is a contravariant functor from the category $\textbf{\textit{Vect}}_{\Bbbk}$ to itself. The particular vector space $\Bbbk$ is self-dual in the sense that there is a canonical identification $\operatorname{Hom}(\Bbbk, \Bbbk) \simeq \Bbbk$:

every linear functional $\Lambda : \mathbb{k} \to \mathbb{k}$ is multiplication by an element $\lambda \in \mathbb{k}$. We identify Λ with λ.

2.1.6 'Hahn–Banach' Lemma. *Given a nonzero vector v in a vector space V, then there exists a linear functional Λ such that $v\Lambda \neq 0$.*

2.1.7 Corollary. *The following natural linear map is injective:*

$$\begin{aligned} V &\to V^{**} = \mathrm{Hom}(V^*, \mathbb{k}) \\ t &\mapsto T := [\Lambda \mapsto t\Lambda]. \end{aligned}$$

Indeed, suppose T is the zero map, i.e. that $t\Lambda = 0$ for all $\Lambda \in V^*$. Then by the lemma, $t = 0$.

2.1.8 Reflexivity of finite-dimensional vector spaces. Suppose V is of finite dimension n, and fix a basis $\{t_1, \ldots, t_n\}$. Let Λ_i be the linear functional which takes value 1 on t_i and zero on the other basis vectors. Then $\{\Lambda_1, \ldots, \Lambda_n\}$ is a basis for V^* called the *dual basis*. In particular, V^* is of dimension n, and therefore isomorphic to V. However, there is no canonical isomorphism.

Consider now the second dual V^{**}. This space is again of dimension n, so the canonical injective map of 2.1.7 is an isomorphism,

$$V \xrightarrow{\sim} V^{**}.$$

(And in the notation of the corollary, $\{T_1, \ldots, T_n\}$ is the dual basis of $\{\Lambda_1, \ldots, \Lambda_n\}$.)

2.1.9 Pairings of vector spaces. A bilinear pairing – or just a *pairing* – of two vector spaces V and W is by definition a linear map $\beta : V \otimes W \to \mathbb{k}$. When we want to write what it does on elements it is convenient to write it

$$\begin{aligned} \beta : V \otimes W &\longrightarrow \mathbb{k} \\ v \otimes w &\longmapsto \langle\, v \,|\, w \,\rangle\,. \end{aligned}$$

2.1.10 Nondegenerate pairings. A pairing $\beta : V \otimes W \to \mathbb{k}$ is called *non-degenerate in the variable* V if there exists a linear map $\gamma : \mathbb{k} \to W \otimes V$, called a *copairing*, such that the following composite is equal to the identity map of V:

$$\begin{array}{ccccc} V & & (V \otimes W) \otimes V & \xrightarrow{\beta \otimes \mathrm{id}_V} & \mathbb{k} \otimes V \\ \| & & \| & & \| \\ V \otimes \mathbb{k} & \xrightarrow{\mathrm{id}_V \otimes \gamma} & V \otimes (W \otimes V) & & V \end{array}$$

Similarly, β is called nondegenerate in the variable W if there exists a copairing $\gamma : \Bbbk \to W \otimes V$, such that the following composite is equal to the identity map of W:

$$\begin{array}{ccccc} \Bbbk \otimes W & \xrightarrow{\gamma \otimes \mathrm{id}_W} & (W \otimes V) \otimes W & & W \\ \| & & \| & & \| \\ W & & W \otimes (V \otimes W) & \xrightarrow{\mathrm{id}_W \otimes \beta} & W \otimes \Bbbk \end{array}$$

These two notions are provisory (but convenient for Lemma 2.1.12 below); the important notion is this: the pairing $\beta : V \otimes W \to \Bbbk$ is simply called *nondegenerate* if it is simultaneously nondegenerate in V and in W.

2.1.11 Lemma. *In that case the two copairings (which were both denoted γ) automatically agree.*

Proof. Let γ_W denote the copairing which makes β nondegenerate in W, and let γ_V denote the copairing which makes β nondegenerate in V. In other words, we have $(\gamma_W \otimes \mathrm{id}_W)(\mathrm{id}_W \otimes \beta) = \mathrm{id}_W$ and $(\mathrm{id}_V \otimes \gamma_V)(\beta \otimes \mathrm{id}_V) = \mathrm{id}_V$. Now consider the composite λ defined as

$$\Bbbk \xrightarrow{\gamma_W \otimes \gamma_V} W \otimes V \otimes W \otimes V \xrightarrow{\mathrm{id}_W \otimes \beta \otimes \mathrm{id}_V} W \otimes V.$$

Factoring λ like this:

$$\Bbbk \xrightarrow{\gamma_V} W \otimes V \xrightarrow{\gamma_W \otimes \mathrm{id}_W \otimes \mathrm{id}_V} W \otimes V \otimes W \otimes V \xrightarrow{\mathrm{id}_W \otimes \beta \otimes \mathrm{id}_V} W \otimes V$$

and using the nondegeneracy in W we see that λ is equal to γ_V. Factoring λ like this:

$$\Bbbk \xrightarrow{\gamma_W} W \otimes V \xrightarrow{\mathrm{id}_W \otimes \mathrm{id}_V \otimes \gamma_W} W \otimes V \otimes W \otimes V \xrightarrow{\mathrm{id}_W \otimes \beta \otimes \mathrm{id}_V} W \otimes V$$

and using the nondegeneracy in V we see that λ is also equal to γ_W. (A neat graphical version of this proof is given in 2.3.23.) □

Let there be a pairing $\beta : V \otimes W \to \Bbbk$. For each fixed second argument $w \in W$ we get a linear functional

$$\begin{aligned} \beta_w : V &\longrightarrow \Bbbk \\ v &\longmapsto \langle v | w \rangle . \end{aligned}$$

Since we have linearity also in the first argument, this actually defines a linear map (the adjoint)

$$\begin{aligned} \beta_{\mathrm{left}} : W &\longrightarrow V^* \\ w &\longmapsto \beta_w = \langle \_ | w \rangle . \end{aligned}$$

Similarly we could fix $v \in V$ to get a linear functional ${}_v\beta : W \to \Bbbk$ (which takes $w \mapsto \langle v \mid w \rangle$), and this defines another linear map

$$\begin{aligned}\beta_{\text{right}} : V &\longrightarrow W^* \\ v &\longmapsto {}_v\beta = \langle v \mid \_ \rangle .\end{aligned}$$

There is a weaker notion of nondegeneracy which is found in many books: $\beta : V \otimes W \to \Bbbk$ is nondegenerate if the two maps β_{right} and β_{left} are injective. We will spend a little while comparing the two definitions. They turn out to agree for finite-dimensional spaces.

2.1.12 Lemma. *The pairing $\beta : V \otimes W \to \Bbbk$ is nondegenerate in W if and only if W is finite-dimensional and the induced map $\beta_{\text{left}} : W \to V^*$ is injective. (Similarly, nondegeneracy in V is equivalent to finite dimensionality of V plus injectivity of $V \to W^*$.)*

Proof. Suppose β is nondegenerate in W; then the copairing $\gamma : \Bbbk \to W \otimes V$ singles out a vector in $W \otimes V$, say $1_\Bbbk \mapsto \sum_{i=1}^{n} w_i \otimes v_i$, for some vectors $w_i \in W$ and $v_i \in V$. Now take an arbitrary $x \in W$ and send it through the composite $W \to W \otimes V \otimes W \to W$:

$$x \mapsto \sum_{i=1}^{n} w_i \otimes v_i \otimes x \mapsto \sum_{i=1}^{n} w_i \langle v_i \mid x \rangle .$$

Nondegeneracy in W means that this composite is the identity map, so we have $x = \sum_{i=1}^{n} w_i \langle v_i \mid x \rangle$ for every $x \in W$. In particular, the vectors $w_1, \ldots, w_n$ span W, which is therefore of finite dimension. Now for the injectivity of

$$\begin{aligned}\beta_{\text{left}} : W &\longrightarrow V^* \\ x &\longmapsto \langle \_ \mid x \rangle :\end{aligned}$$

Suppose $\langle \_ \mid x \rangle$ is the zero functional. Then in particular for the vectors $v_1, \ldots, v_n$ we have $\langle v_i \mid x \rangle = 0$. But these scalars are exactly the coordinates of x in the 'basis' $w_1, \ldots, w_n$, so in particular x is zero itself. This shows that $W \to V^*$ is injective.

Conversely, suppose that W is of finite dimension (so we can choose a basis $w_1, \ldots, w_n$), and that $W \to V^*$ is injective. Since the w_j are linearly independent, injectivity implies that the functionals $\langle \_ \mid w_j \rangle$ are also linearly independent. Then there exist vectors $v_1, \ldots, v_n$ such that $\langle v_i \mid w_i \rangle = 1$ and $\langle v_i \mid w_j \rangle = 0$ for $i \neq j$.

Define the copairing γ by setting $1_{\Bbbk} \mapsto \sum_i w_i \otimes v_i$. Now send a general vector $\sum_{j=1}^{n} w_j \lambda_j$ through the composite $W \to W \otimes V \otimes W \to W$:

$$\sum_{j=1}^{n} w_j \lambda_j \mapsto \sum_{i,j} w_i \otimes v_i \otimes w_j \lambda_j \mapsto \sum_{i,j} w_i \lambda_j \langle v_i \,|\, w_j \rangle = \sum_{i=1}^{n} w_i \lambda_i,$$

so β is nondegenerate in W as claimed. □

2.1.13 Example. The obvious evaluation pairing

$$\begin{aligned} V \otimes V^* &\longrightarrow \Bbbk \\ v \otimes \Lambda &\longmapsto v\Lambda \end{aligned}$$

is not always nondegenerate in our strong sense of the word. It is nondegenerate if and only if V is of finite dimension. This follows from 2.1.12 and 2.1.7.

2.1.14 Lemma. *Let V and W be vector spaces of finite dimension, and consider a pairing $\beta : V \otimes W \to \Bbbk$ as above. Then β_{right} is the dual map of β_{left} (modulo the identification $V \xrightarrow{\sim} V^{**}$ of 2.1.8). Also, β_{left} is the dual map of β_{right} (modulo the identification $W \xrightarrow{\sim} W^{**}$).*

Proof. By definition the dual of β_{left} is

$$\begin{aligned} \operatorname{Hom}(V^*, \Bbbk) &\longrightarrow \operatorname{Hom}(W, \Bbbk) \\ T &\longmapsto \beta_{\text{left}}\, T \end{aligned}$$

sending the map $V^* \xrightarrow{T} \Bbbk$ to the composite $W \xrightarrow{\beta_{\text{left}}} V^* \xrightarrow{T} \Bbbk$. We already explained the identification $V \xrightarrow{\sim} \operatorname{Hom}(V^*, \Bbbk)$, $t \mapsto T := [\Lambda \mapsto t\Lambda]$. Let us now write out the details of the composite

$$\begin{aligned} V &\to \operatorname{Hom}(W, \Bbbk) \\ t &\mapsto \beta_{\text{left}}\, T \end{aligned}$$

to see what the linear form $\beta_{\text{left}}\, T$ does to an element $z \in W$. Well, $z\beta_{\text{left}}$ is the linear form β_z, and applying T to it means evaluating it at t, which by definition is $\langle t \,|\, z \rangle$. So in conclusion, the dual of β_{left} is the map that takes t to $[z \mapsto \langle t \,|\, z \rangle]$. This is exactly β_{right}. □

For finite-dimensional vector spaces, the dualising functor is a (contravariant) equivalence of categories, so in particular it preserves the property of being invertible. So in this case β_{left} is injective if and only if β_{right} is. Thus,

2.1.15 Lemma. *Given a pairing*

$$\begin{aligned} \beta : V \otimes W &\longrightarrow \Bbbk \\ v \otimes w &\longmapsto \langle\, v \,|\, w \,\rangle\,, \end{aligned}$$

between finite-dimensional vector spaces, the following are equivalent.

(i) *β is nondegenerate.*
(ii) *The induced linear map $\beta_{\text{left}} : W \to V^*$ is an isomorphism.*
(iii) *The induced linear map $\beta_{\text{right}} : V \to W^*$ is an isomorphism.*

In that case, clearly all the four involved vector spaces are of the same dimension. On the other hand, if we already know for other reasons that V and W are of the same dimension, then 'being an isomorphism' in the lemma is equivalent to just being injective, so in that case, nondegeneracy can also be characterised by each of the following a priori weaker conditions:

(ii′) $\langle\, v \,|\, w \,\rangle = 0\ \forall v \in V \ \Rightarrow\ w = 0$
(iii′) $\langle\, v \,|\, w \,\rangle = 0\ \forall w \in W \ \Rightarrow\ v = 0.$

This is perhaps the most usual definition of nondegeneracy...

2.1.16 Remark. It should be noted that in terms of coordinates (in the finite-dimensional situation), nondegeneracy just amounts to saying that the matrix expressing β (and β_{left} and β_{right}) is invertible, cf. 2.3.34. (Nondegeneracy in one variable amounts to having a left inverse, and in the other variable means existence of a right inverse; these two must of course agree – this proves Lemma 2.1.11.) For more details on this viewpoint, see the subsection on coordinates (page 123).

2.1.17 Duals of tensor products. *Let V and W be $\Bbbk$-vector spaces of finite dimension. Then the canonical linear map*

$$\begin{aligned} W^* \otimes V^* &\longrightarrow (V \otimes W)^* \\ \psi \otimes \phi &\longmapsto \big[x \otimes y \mapsto (x\phi)(y\psi)\big] \end{aligned}$$

is an isomorphism.

Note that there is a twist on the factors when we take the dual of a tensor product. This may seem strange at first, but notice in the proof how it comes about naturally. We will also see in 2.1.36 that the twist is necessary in order to get natural properties for A-modules.

Proof. Set up a pairing

$$\rho \,:\, (V \otimes W) \otimes (W^* \otimes V^*) \longrightarrow \Bbbk$$

defined by first coupling the two middle modules $W \otimes W^* \to \Bbbk$, and then coupling the two outer ones. In other words, it is the composite

$$V \otimes W \otimes W^* \otimes V^* \to V \otimes \Bbbk \otimes V^* \to V \otimes V^* \to \Bbbk,$$

where the first and the last maps are the canonical evaluation pairings 2.1.13 (and the middle map is the canonical identification). The induced linear map $\rho_{\text{left}} : W^* \otimes V^* \to \big(V \otimes W\big)^*$ is precisely the map of the statement. The two evaluation pairings are nondegenerate since we are in the finite-dimensional case, so there are corresponding copairings $\Bbbk \to W^* \otimes W$ and $\Bbbk \to V^* \otimes V$ satisfying the identities of 2.1.10. Use these copairings to define

$$\Bbbk \to W^* \otimes W \to W^* \otimes \Bbbk \otimes W \to W^* \otimes V^* \otimes V \otimes W$$

which is easily seen to be a copairing for ρ satisfying the conditions 2.1.10, so ρ is nondegenerate. Then by Lemma 2.1.15, the induced linear map ρ_{left} is an isomorphism. □

Algebras and modules

2.1.18 $\Bbbk$-algebras. A $\Bbbk$-*algebra* is a $\Bbbk$-vector space A together with two $\Bbbk$-linear maps

$$\mu : A \otimes A \to A, \qquad \eta : \Bbbk \to A$$

(called multiplication and unit map) such that these three diagrams commute:

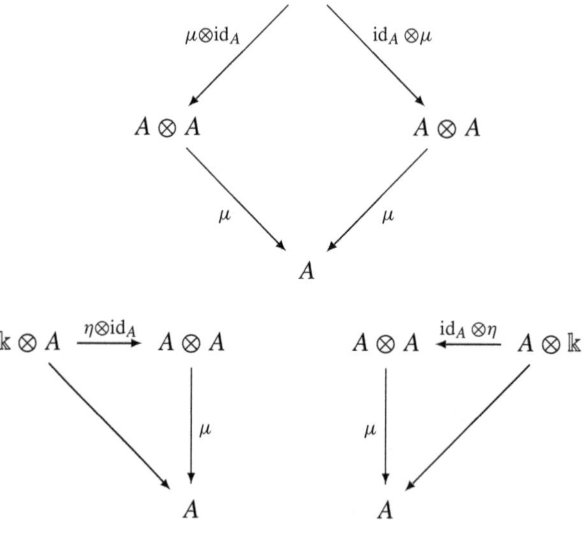

The symbols id_A stand for the identity linear map $A \to A$, and the diagonal maps without labels are scalar multiplication which are canonical isomorphisms.

In terms of elements, we will write the multiplication by plain juxtaposition:

$$\begin{aligned} A \otimes A &\longrightarrow A \\ x \otimes y &\longmapsto xy \end{aligned}$$

and we let 1_A (or simply 1) denote the image of $1_{\mathbb{k}}$ under the map $\eta : \mathbb{k} \to A$. Then we can write the axioms in terms of elements of A:

$$(xy)z = x(yz) \qquad \text{and} \qquad 1x = x = x1.$$

These conditions are associativity and unity requirements. (Note that we do not require commutativity at this stage.)

2.1.19 Remarks. There are three structures involved in the definition of a $\mathbb{k}$-algebra: the $\mathbb{k}$-structure (allowing multiplication with $\mathbb{k}$-scalars) and two composition laws (addition and multiplication). In the above definition, we first bundled together the $\mathbb{k}$-structure and the addition, stipulating that A be first of all a vector space, and then we imposed the multiplication. By defining the multiplication in terms of tensor products of vector spaces instead of just using the cartesian product we encoded distributivity: having a linear map $A \otimes A \to A$ is equivalent to having a bilinear map $A \times A \to A$, and this in turn is equivalent to saying that the multiplication defined by this map distributes over sums. In particular, A is a *ring*.

2.1.20 Example. The particular vector space $\mathbb{k}$ is canonically a $\mathbb{k}$-algebra: let $\mu_{\mathbb{k}} : \mathbb{k} \otimes \mathbb{k} \to \mathbb{k}$ be multiplication in the field $\mathbb{k}$ (it is an isomorphism), and let $\eta_{\mathbb{k}} : \mathbb{k} \to \mathbb{k}$ be the identity map (also an isomorphism). Clearly the axioms are satisfied. Also it follows that the map $\eta : \mathbb{k} \to A$ is a ring homomorphism, i.e. that this diagram commutes:

$$\begin{array}{ccc} \mathbb{k} \otimes \mathbb{k} & \xrightarrow{\eta \otimes \eta} & A \otimes A \\ \downarrow & & \downarrow \mu \\ \mathbb{k} & \xrightarrow[\eta]{} & A \end{array}$$

2.1.21 Alternative definition of $\mathbb{k}$-algebras. One can also define $\mathbb{k}$-algebras by starting with the two composition laws and then imposing the $\mathbb{k}$-structure, saying: *a $\mathbb{k}$-algebra is a ring A equipped with a ring homomorphism $\mathbb{k} \to A$.*

(See 3.5.9 for more comments in this vein, using the language of monoids in monoidal categories.)

2.1.22 $\Bbbk$-algebra homomorphisms. Of course there is also a notion of $\Bbbk$-*algebra homomorphism*: maps that preserve the structure. The reader can copy the exact definition from 3.1.6. Let $\mathbf{Alg}_{\Bbbk}$ denote the category of $\Bbbk$-algebras and $\Bbbk$-algebra homomorphisms. (We will mostly be concerned with $\Bbbk$-algebras isolatedly, so we will not need the notion of $\Bbbk$-algebra homomorphisms very much.)

2.1.23 Right A-modules. A *right A-module* is a vector space M together with a $\Bbbk$-linear map (called a right action of A on M)

$$\alpha : M \otimes A \to M$$

which satisfies the axioms expressed by the commutativity of the two diagrams

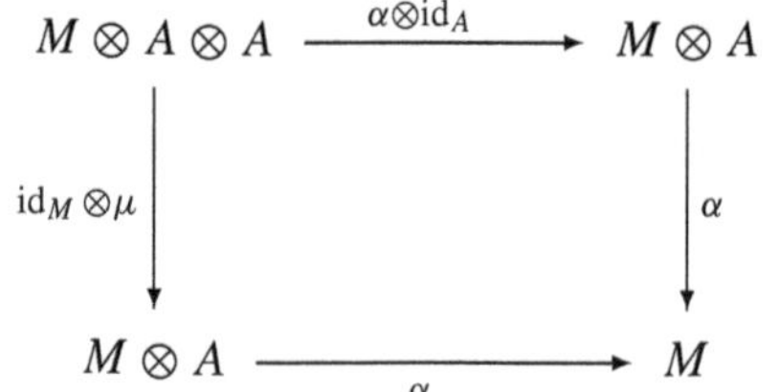

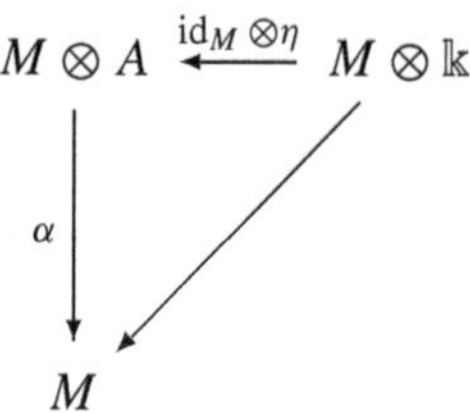

If we write the action like this:

$$\begin{aligned} M \otimes A &\xrightarrow{\alpha} M \\ x \otimes a &\longmapsto x.a \end{aligned}$$

then we can write the axioms in terms of elements:

$$(x.a).b = x.(ab) \qquad \text{and} \qquad x.1 = x.$$

2.1.24 Note again that distributivity is encoded in the tensor product.

2.1.25 Example. Let A be a $\Bbbk$-algebra. Then A is of course naturally a right A-module itself, since we have the multiplication map $\mu : A \otimes A \to A$ as a special case of the action. The associativity and the unit axioms for μ can then be regarded as special instances of the two axioms for the action.

2.1.26 Right A-homomorphisms. Let M and N be right A-modules. A $\Bbbk$-linear map $\phi : M \to N$ is called a *right A-homomorphism* if this diagram commutes:

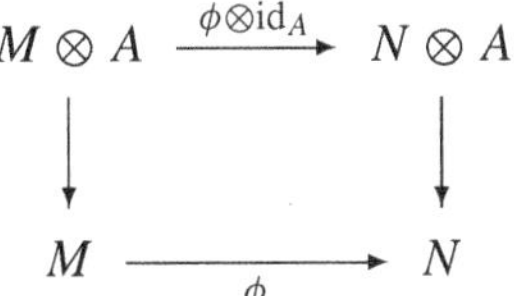

In other words, for all $x \in M$ and all $a \in A$ we have $(x.a)\phi = (x\phi).a$. We also say that ϕ is *right A-linear.*

(It is clear that the identity map of a right A-module is right A-linear, and that the composition of two right A-homomorphisms is again a right A-homomorphism, so the right A-modules and the right A-homomorphisms form a category denoted $\textbf{\textit{rMod}}_A$.)

2.1.27 Left A-modules and left A-homomorphisms are defined similarly: a *left A-module* is a vector space M with a $\Bbbk$-linear map $A \otimes M \to M$ (written $a \otimes x \mapsto a.x$), satisfying axioms similar to those of right A-modules, $a.(b.x) = (ab).x$ and $1.x = x$. (If A is commutative, then the two notions of A-modules coincide.) A $\Bbbk$-linear map $\phi : M \to N$ between two left A-modules is called a *left A-homomorphism* if it satisfies $(a.x)\phi = a.(x\phi)$ for all $x \in M$ and all $a \in A$. (This looks neater than the formula for right A-linearity 2.1.26, since there is no 'changing place' involved. The asymmetry stems from the fact that we always write functions on the right-hand side of their argument, independent of left or right structure – with traditional 'functions-at-left' notation we would get the opposite asymmetry: the right-linearity condition would look better than the left-linearity condition.) In fact there is complete symmetry between the two notions (as one sees clearly by writing out the axioms in diagrams as on page 86), and everything that holds for right A-modules will have a left A version as well. The reason why we have to bother with both sorts of modules is duality:

2.1.28 The dual of a module. Suppose M is a right A-module. Then the dual vector space $M^* := \mathrm{Hom}(M, \Bbbk)$ has a canonical left A-module structure given by:

$$\begin{aligned} A \otimes M^* &\longrightarrow M^* \\ a \otimes \Lambda &\longmapsto a.\Lambda := [x \mapsto (x.a)\Lambda]. \end{aligned}$$

Similarly, if M is a left A-module, then M^* becomes a right A-module via the rule $x(\Lambda.a) = (a.x)\Lambda$ (with $\Lambda \in M^*, a \in A, x \in M$).

2.1.29 Dual maps. *Let M and N be right A-modules, and let $\psi : M \to N$ be a right A-homomorphism. Then the dual map*

$$\begin{aligned} \psi^* : N^* &\to M^* \\ \Lambda &\mapsto \psi\,\Lambda \end{aligned}$$

is a left A-homomorphism.

Proof. Right A-linearity (cf. 2.1.26) of ψ means that for $a \in A$ and $x \in M$ we have $(x.a)\psi = (x\psi).a$. We must check the commutativity of the square

$$\begin{array}{ccc} A \otimes N^* & \xrightarrow{\mathrm{id}_A \otimes \psi^*} & A \otimes M^* \\ \downarrow & & \downarrow \\ N^* & \xrightarrow[\psi^*]{} & M^* \end{array}$$

The upper way around, an $a \otimes \Lambda$ is first sent to $a \otimes \psi\Lambda$ and then to $a.(\psi\Lambda)$. This is the linear form on M given by $x \mapsto x.a \mapsto (x.a)\psi\Lambda$. On the other hand, taking the lower way around the diagram we find $a \otimes \Lambda \mapsto a.\Lambda \mapsto \psi(a.\Lambda)$ (which is the composition of first ψ, then $a.\Lambda$). This form goes $x \mapsto x\psi \mapsto (x\psi)(a.\Lambda) = ((x\psi).a)\Lambda = (x.a)\psi\Lambda$, by the linearity of ψ. □

In this way, taking the dual on modules and maps is a contravariant functor from the category of right A-modules to the category of left A-modules. Assuming that all modules are of finite dimension over $\Bbbk$, then this functor is a contravariant equivalence of categories.

Similarly if M and N are left A-modules and $\psi : M \to N$ is left A-linear, then precomposition with ψ provides a right A-module homomorphism $N^* \to M^*$ (and a contravariant functor from left A-modules to right A-modules).

2.1.30 Reflexivity. *If M is a right A-module of finite dimension over $\Bbbk$, then the vector space identification $M \xrightarrow{\sim} M^{**}$ is a right A-isomorphism (and if M is a left A-module then the identification is a left A-isomorphism).*

Proof. As in 2.1.8, for fixed $t \in M$ let T denote the element in M^{**} which is evaluation at t. So it is: $\Lambda \mapsto t\Lambda$. Since M^* is a left A-module, M^{**} is a right A-module, so it makes sense to multiply an A-scalar a on T from the right. By definition, $T.a$ is the map $\Lambda \mapsto (a.\Lambda)T = t(a.\Lambda) = (t.a)\Lambda$, in other words, it is simply evaluation at $t.a$. This shows that $M \to M^{**}$ is right A-linear. □

2.1.31 Pairings of A-modules. Given a vector space pairing $\beta : M \otimes N \to \Bbbk$, it has no meaning to ask whether it is A-linear (no matter which A-structures M and N might have) because $\Bbbk$ is not an A-module. But suppose M is a right

A-module and N is a left A-module. Then it is natural to ask whether β_{left} : $N \to M^*$ is left A-linear, and whether $\beta_{\text{right}} : M \to N^*$ is right A-linear.

2.1.32 Associative pairings. A pairing $\beta : M \otimes N \to \Bbbk$ as above is said to be *associative* if this diagram commutes:

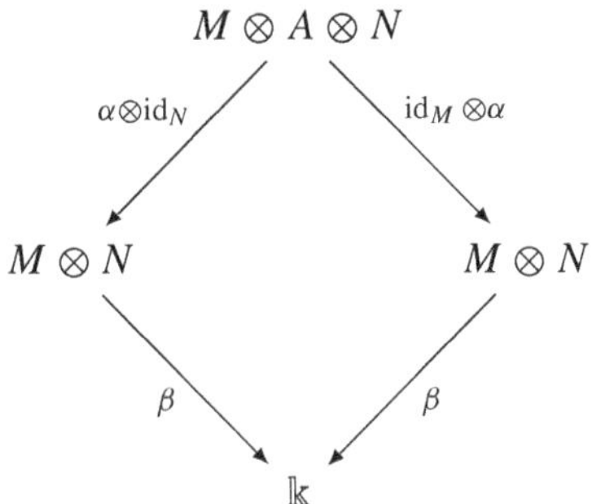

In other words, the pairing $x \otimes y \mapsto \langle x \,|\, y \rangle$ is associative when

$$\langle xa \,|\, y \rangle = \langle x \,|\, ay \rangle \qquad \text{for every } x \in M, a \in A, y \in N.$$

2.1.33 Lemma. *For a pairing $M \otimes N \to \Bbbk$ as above, the following three statements are equivalent:*

(i) $M \otimes N \to \Bbbk$ *is associative,*
(ii) $N \to M^*$ *is left A-linear,*
(iii) $M \to N^*$ *is right A-linear.*

Proof. Since associativity is a symmetric condition it is enough to show (i)$\Leftrightarrow$(ii). Consider the diagram expressing left A-linearity of $N \to M^*$:

$$\begin{array}{ccc} A \otimes N & \xrightarrow{\mathrm{id}_A \otimes \beta_{\text{left}}} & A \otimes M^* \\ \downarrow & & \downarrow \\ N & \xrightarrow[\beta_{\text{left}}]{} & M^* \end{array}$$

Going the upper way around, $a \otimes y$ is sent to the linear form $x \mapsto \langle xa \,|\, y \rangle$. Going the lower way we arrive at the form $x \mapsto \langle x \,|\, ay \rangle$. Commutativity of the diagram means that these two expressions are equal for all values of x, a, y, so it is equivalent to the associativity condition. □

2.1.34 Example. If M is a right A-module then the obvious evaluation pairing

$$\begin{aligned} M \otimes M^* &\longrightarrow \Bbbk \\ x \otimes \Lambda &\longmapsto x\Lambda \end{aligned}$$

is associative, by definition of the left A-structure on M^*.

2.1.35 A-module structure on tensor products. Let M and N be $\Bbbk$-vector spaces, and consider the tensor product $M \otimes N$. If M is also a left A-module, then $M \otimes N$ acquires left A-module structure by the obvious map (multiplication on M from the left)

$$A \otimes M \otimes N \to M \otimes N.$$

If N is a right A-module we similarly get a right A-structure on the tensor product by

$$M \otimes N \otimes A \to M \otimes N.$$

2.1.36 Duals of tensor products. *Let N be a right A-module, and let M be a vector space. Then the canonical linear map (2.1.17)*

$$\begin{aligned} N^* \otimes M^* &\longrightarrow \big(M \otimes N\big)^* \\ \psi \otimes \phi &\longmapsto \big[x \otimes y \mapsto (x\phi)(y\psi)\big] \end{aligned}$$

is left A-linear.

Proof. Consider the pairing

$$(M \otimes N) \otimes (N^* \otimes M^*) \longrightarrow \Bbbk$$

defined by first coupling the two middle modules $N \otimes N^* \to \Bbbk$, and then coupling the two outer ones. This pairing is associative because the pairing $N \otimes N^*$ is so (cf. 2.1.34). But by Lemma 2.1.33 this is equivalent to left A-linearity of the adjoint map $N^* \otimes M^* \to \big(M \otimes N\big)^*$, which is precisely the map of the assertion.

(Note that it does not matter whether M has any A-structure.) □

Exercises

1. Let $S = \{x_1, \ldots, x_n\}$ be a set, and consider the vector space $\Bbbk S = \{\sum_i a_i x_i \mid a_i \in \Bbbk\}$ of all formal linear combinations of the elements of S. Show that $S \mapsto \Bbbk S$ defines a functor $\textbf{\textit{FinSet}} \to \textbf{\textit{Vect}}_\Bbbk$. In the other direction there is the forgetful functor that to a vector space W associates the underlying set, denoted $|W|$. Show that for any finite set S, and for any vector space W there is a bijection

$$\textbf{\textit{Vect}}_\Bbbk(\Bbbk S, W) \leftrightarrow \textbf{\textit{Set}}(S, |W|).$$

2. Suppose S and T are finite sets, and there is an isomorphism of vector spaces $\Bbbk S \xrightarrow{\sim} \Bbbk T$. Show that S and T have the same cardinality.
3. Provide the details of the argument in Remark 2.1.16. Let V be a vector space of finite dimension equipped with a pairing $\langle \mid \rangle$. Fix a basis $\{T_1, \ldots, T_n\}$ and consider the matrix whose ijth entry is $\langle T_i \mid T_j \rangle$. Show that the pairing is nondegenerate if and only if the matrix is invertible.
4. Show that the evaluation pairing $V \otimes V^* \to \Bbbk$ of a finite-dimensional vector space has this universal property: for any nondegenerate pairing $\beta : V \otimes W \to \Bbbk$ there is a unique isomorphism $W \xrightarrow{\sim} V^*$ compatible with the two pairings.
5. Let V' and V'' be vector spaces, equipped with linear functionals $\varepsilon' : V' \to \Bbbk$ and $\varepsilon'' : V'' \to \Bbbk$. Show that

$$\begin{aligned} V' \oplus V'' &\longrightarrow \Bbbk \\ (v', v'') &\longmapsto v'\varepsilon' + v''\varepsilon'' \end{aligned}$$

defines a linear functional on the vector space $V' \oplus V''$.
6. Show that if $M \to N$ is a right A-homomorphism which is invertible as $\Bbbk$-linear map, then the inverse is also a right A-homomorphism.
7. Let $S = \{x_1, \ldots, x_n\}$ be a finite set and consider the algebra of all polynomials in the indeterminates $x_1, \ldots, x_n$, with $\Bbbk$-coefficients. Show that this construction defines a functor ***FinSet*** $\to$ $\boldsymbol{Alg}_{\Bbbk}$. Under what condition on the set S is the image a finite-dimensional algebra?
8. Direct product of algebras. Let A' and A'' be $\Bbbk$-algebras. Show that the direct sum vector space $A := A' \oplus A''$ becomes a $\Bbbk$-algebra under coordinatewise multiplication. Usually this algebra is denoted $A' \times A'$ and called the direct product, but as long as we emphasise the vector space structure (e.g. look at subspaces) we continue to write $\oplus$.
   Show that $e' := 1 \oplus 0$ is a central idempotent in A, and that $A' \oplus 0$ is a two-sided ideal in A (but not a subalgebra) – similarly of course for $e'' := 0 \oplus 1$ and $0 \oplus A''$. (*Idempotent* means $e'e' = e'$; *central* means $e'a = ae'$ for all $a \in A$.)
   Show that every left ideal in $A' \times A''$ is of the form $\mathfrak{a}' \oplus \mathfrak{a}''$, where $\mathfrak{a}' \subset A'$ and $\mathfrak{a}'' \subset A''$ are left ideals.
   Conversely, if A is any $\Bbbk$-algebra, and $e' \in A$ is a central idempotent (then so is $e'' := 1 - e'$); show that A decomposes into the direct product of two algebras, i.e. is isomorphic to some $A' \oplus A''$ in such a way that $1 \oplus 0$ corresponds to e' and $0 \oplus 1$ corresponds to e''.

9. Recall that a ring A is *Artinian* if every descending chain of (left) ideals $\mathfrak{a}_0 \supset \mathfrak{a}_1 \supset \dots$ is stationary (i.e. there is a k_0 such that for $k \geq k_0$ we have $\mathfrak{a}_k = \mathfrak{a}_{k+1}$). Show that if A is a finite-dimensional $\Bbbk$-algebra then it is Artinian.

2.2 Definition and examples of Frobenius algebras

Definition and basic properties

Now we specialise all our constructions to the case where the A-modules in question are simply the $\Bbbk$-algebra A itself. Throughout this section we assume A to be of finite dimension. Naturally A comes with both a left and a right A-structure.

A linear functional $\Lambda : A \to \Bbbk$ defines a hyperplane in A,

$$\mathrm{Null}(\Lambda) := \{x \in A \mid x\Lambda = 0\}.$$

We call it the *nullspace* rather than the kernel, just to remind us that it is not an ideal or subalgebra in A, but merely a linear subspace.

2.2.1 Definition of Frobenius algebra. A *Frobenius algebra* is a $\Bbbk$-algebra A of finite dimension, equipped with a linear functional $\varepsilon : A \to \Bbbk$ whose nullspace contains no nontrivial left ideals. The functional $\varepsilon \in A^*$ is called a *Frobenius form*.

2.2.2 Remarks. The Frobenius form is part of the structure. We will see in 2.2.7 that a given algebra may allow various distinct Frobenius forms. Equivalent characterisations of Frobenius algebras will be given shortly (in 2.2.5 and 2.2.6) – among other things we will see that we could equally well have stated the definition in terms of right ideals.

Let us remark that having no nontrivial left ideals in $\mathrm{Null}(\varepsilon)$ is equivalent to having no nontrivial *principal* left ideals in $\mathrm{Null}(\varepsilon)$, since every nonzero left ideal contains a nonzero principal left ideal – simply take a nonzero element of it and let that element generate a principal ideal. So the condition can be phrased like this

$$(Ay)\varepsilon = 0 \ \Rightarrow \ y = 0.$$

2.2.3 Functionals and associative pairings on A. Every linear functional $\varepsilon : A \to \Bbbk$ (Frobenius or not) determines canonically a pairing $A \otimes A \to \Bbbk$, namely $x \otimes y \mapsto (xy)\varepsilon$. Clearly this pairing is associative (cf. Definition 2.1.32). Conversely, given an associative pairing $A \otimes A \to \Bbbk$, denoted $x \otimes y \mapsto \langle x \,|\, y \rangle$, a linear functional is canonically determined, namely

$$\begin{aligned} A &\longrightarrow \Bbbk \\ a &\longmapsto \langle 1_A \,|\, a \rangle = \langle a \,|\, 1_A \rangle . \end{aligned}$$

You can easily check that if you let one construction follow the other you get back what you started with, so there is a one-to-one correspondence between linear functionals on A and associative pairings.

Now the important remark is:

2.2.4 Lemma. *Let $\varepsilon : A \to \Bbbk$ be a linear functional and let $\langle \ | \ \rangle$ denote the corresponding associative pairing $A \otimes A \to \Bbbk$. Then the following are equivalent.*

(i) *The pairing is nondegenerate.*
(ii) Null(ε) *contains no nontrivial left ideals.*
(iii) Null(ε) *contains no nontrivial right ideals.*

In particular this shows that in the definition of Frobenius algebra we could have used right ideals instead of left ideals.

Proof. We have already done the work. Recall from 2.1.15 (ii′) that $\langle \ | \ \rangle$ is nondegenerate if and only if

$$\langle A \,|\, y \rangle = 0 \;\Rightarrow\; y = 0.$$

(Note that this condition (injectivity) is sufficient since the paired modules A and A are of the same dimension!) On the other hand, by the way ε and $\langle \ | \ \rangle$ determine each other this just means

$$(Ay)\varepsilon = 0 \;\Rightarrow\; y = 0,$$

which in turn is equivalent to saying that Null(ε) contains no left ideals. This proves (i)⇔(ii). The equivalence (i)⇔(iii) follows by using the 'nondegeneracy in the other variable' instead (cf. 2.1.15 (iii′)). □

Since the data of an associative bilinear pairing and a linear functional completely determine each other as above, we can give the following.

2.2.5 Alternative definition of Frobenius algebra. A *Frobenius algebra* is a $\Bbbk$-algebra A of finite dimension, equipped with an associative nondegenerate pairing $\beta : A \otimes A \to \Bbbk$. We call this pairing the *Frobenius pairing*.

Lemma 2.2.4 shows that these two definitions are equivalent in the sense that the structure of one definition canonically induces the structure of the other definition.

This second definition of Frobenius algebras quickly leads to a couple of other characterisations, since we have other ways to characterise nondegeneracy of a pairing. Recall from 2.1.15 that given a nondegenerate pairing $\beta : A \otimes A \to \Bbbk$ there are induced two $\Bbbk$-linear isomorphisms

$$\beta_{\text{left}} : A \xrightarrow{\sim} A^*, \qquad \beta_{\text{right}} : A \xrightarrow{\sim} A^*.$$

Furthermore, we showed in 2.1.33 that associativity of β is equivalent to the left A-linearity of β_{left}, and also equivalent to the right A-linearity of β_{right}. Note that even though these two maps have the same source and target there is no reason for them to be equal! (Precisely, they are dual. We will see in a minute under what circumstances they agree.) This leads to

2.2.6 Third definition of Frobenius algebra. A *Frobenius algebra* is a finite-dimensional $\Bbbk$-algebra A equipped with a left A-isomorphism to its dual. Alternatively (and equivalently) A is equipped with a right A-isomorphism to its dual.

The preceding discussion shows how each of these two structures is naturally induced from the Frobenius pairing. Conversely, given a left A-linear isomorphism $A \xrightarrow{\sim} A^*$, we can reconstruct the nondegenerate pairing (and it will be associative because of the left A-linearity of the isomorphism). A right A-linear isomorphism would do as well...

Alternatively we can relate these A-isomorphisms directly to the Frobenius form of our first definition of Frobenius algebra. Given a left A-isomorphism $A \xrightarrow{\sim} A^*$ we get a linear functional which is simply the image of 1_A in A^*. The fact that this functional has no nontrivial left ideals in its nullspace follows readily from the fact that $A \xrightarrow{\sim} A^*$ is injective – and that it is left A-linear. Similarly, a right A-isomorphism $A \xrightarrow{\sim} A^*$ determines naturally a Frobenius form (as the image of 1_A).

Conversely, given the Frobenius form $\varepsilon : A \to \Bbbk$, we construct an isomorphism of left A-modules $A \xrightarrow{\sim} A^*$ by putting $1_A \mapsto \varepsilon$ and extending left A-linearly. This left A-homomorphism is injective since there are no nontrivial left ideals in $\text{Null}(\varepsilon)$. Since furthermore the two spaces have the same dimension over $\Bbbk$, it is also surjective.

2.2.7 About the choice of structure. To recapitulate, given a finite-dimensional $\Bbbk$-algebra A, we have four definitions of Frobenius structure.

- A linear functional $\varepsilon : A \to \Bbbk$ whose nullspace contains no nontrivial left ideals.
- An associative nondegenerate pairing $\beta : A \otimes A \to \Bbbk$.
- A left A-isomorphism $A \xrightarrow{\sim} A^*$.
- A right A-isomorphism $A \xrightarrow{\sim} A^*$.

The four different versions of Frobenius structure are canonically determined by each other, and therefore we think of them as being one and the same structure. But this structure is not uniquely given. For example, if $\varepsilon : A \to \Bbbk$ is a Frobenius form, and $u \in A$ is invertible, then the functional $x \mapsto (xu)\varepsilon$ is also a Frobenius form. Indeed, although the nullspaces of these two functionals are not the same, the ideals they contain must be the same, since u is invertible, so if one is a Frobenius form then the other is as well. In fact, if $\varepsilon : A \to \Bbbk$ is a Frobenius form then every other Frobenius form ε' on A is given in this way. To see this, consider the two induced left A-isomorphisms

$$A \xrightarrow{\sim} A^* \qquad\qquad A \xrightarrow{\sim} A^*$$
$$1 \mapsto \varepsilon \qquad\qquad 1 \mapsto \varepsilon'$$

then there is a unique left A-linear comparison homomorphism $A^* \leftrightarrow A^*$ (one in each direction); since everything is left A-linear, these two maps are given by right multiplication by elements u and u' in A, so that $\varepsilon' = u.\varepsilon$ and $\varepsilon = u'.\varepsilon'$; clearly u and u' are inverses of each other in A.

Let us record this in a lemma:

2.2.8 Lemma. *If A is a $\Bbbk$-algebra with Frobenius form ε, then every other Frobenius form on A is given by precomposing ε with multiplication by an invertible element of A. Equivalently, given a fixed left A-isomorphism $\theta : A \xrightarrow{\sim} A^*$, then the elements in A^* which are Frobenius forms are precisely the images of the invertible elements in A.* □

Before coming to the examples, we should mention an important class of Frobenius algebras.

2.2.9 Symmetric Frobenius algebras. A Frobenius algebra A is called a *symmetric Frobenius algebra* if one (and hence all) of the following equivalent conditions holds.

(i) The Frobenius form $\varepsilon : A \to \Bbbk$ is *central*; this means that $(ab)\varepsilon = (ba)\varepsilon$ for all $a, b \in A$.
(ii) The pairing $\langle \ | \ \rangle$ is symmetric (i.e. $\langle a | b \rangle = \langle b | a \rangle$ for all $a, b \in A$).
(iii) The left A-isomorphism $A \xrightarrow{\sim} A^*$ is also right A-linear.
(iv) The right A-isomorphism $A \xrightarrow{\sim} A^*$ is also left A-linear.

In fact, in this case, since the two maps $A \xrightarrow{\sim} A^*$ agree on 1_A, they coincide. (The proof is left as an exercise.)

Clearly a commutative Frobenius algebra is always symmetric. Symmetric Frobenius algebras are often plainly called *symmetric algebras* (cf. Curtis and Reiner [15]). In Quinn [43], they are called ambialgebras.

The condition $(ab)\varepsilon = (ba)\varepsilon$ characterising central (Frobenius) forms is often referred to as the *trace condition* – see 2.2.16.

2.2.10 Important remarks. In our setting, being a Frobenius algebra is a structure, not a property. This means that given an algebra it does not make sense to ask whether it is Frobenius or not; the Frobenius form must be specified. We often abuse language, saying 'let A be a Frobenius algebra', without specifying the Frobenius structure (Frobenius form or Frobenius pairing); when we do this we are tacitly assuming that a particular structure has been chosen (or worse, we might just intend to say that a Frobenius form exists). We ought to say: 'let (A, ε) be a Frobenius algebra'.

Being symmetric is a property, not a structure: this means that given a Frobenius algebra (that is, the algebra together with its Frobenius structure) then it makes sense to ask whether it is symmetric or not; it is not something you can choose.

As an illustration of these considerations, suppose A is a $\Bbbk$-algebra, and that $\varepsilon : A \to \Bbbk$ and $\varepsilon' : A \to \Bbbk$ are two Frobenius structures (so we are talking about two different Frobenius algebras, even though the underlying algebra A is the same). Then it can easily happen that (A, ε) is symmetric while (A, ε') is not!

Precisely we have this lemma (whose proof is an exercise):

2.2.11 Lemma. *Let (A, ε) be a symmetric Frobenius algebra (i.e. ε is central); then every other central Frobenius form on A is given by multiplication with a central invertible element of A.*

Recall that an element of a ring is called central if it commutes with every other element.

Examples

Here is a collection of examples of Frobenius algebras. Although some of them are quite advanced compared to the level of prerequisites assumed elsewhere in this book, hopefully the reader will enjoy seeing the animals in their natural habitat...

In each example, A is assumed to be a $\Bbbk$-algebra of finite dimension over $\Bbbk$.

2.2.12 The trivial Frobenius algebra. Let $A = \Bbbk$, and let $\varepsilon : A \to \Bbbk$ be the identity map of $\Bbbk$. Clearly there are no ideals in the kernel of this map, so we have a Frobenius algebra.

2.2.13 Algebraic field extensions. Let A be a finite field extension of $\Bbbk$. Since fields have no nontrivial ideals, any nonzero $\Bbbk$-linear map $A \to \Bbbk$ will do as Frobenius form. For the expert: if A is *separable* over $\Bbbk$ then the trace map $A \to \Bbbk$ is a natural choice of Frobenius form (see Lang [30] for these notions).

2.2.14 Concrete example. The field of complex numbers $\mathbb{C}$ is a Frobenius algebra over $\mathbb{R}$: an obvious Frobenius form is 'taking the real part'

$$\begin{aligned} \mathbb{C} &\longrightarrow \mathbb{R} \\ a + ib &\longmapsto a. \end{aligned}$$

(But it could also be something more exotic, like $2 + 3i \mapsto 7$, $1 - i \mapsto 4$ (then by linear algebra, the line spanned by $1 + 19i$ maps to 0).)

2.2.15 Skew-fields. Let A be a skew-field (also called division algebra) of finite dimension over $\Bbbk$. Since just like a field, a skew-field has no nontrivial left ideals (or right ideals), any nonzero linear form $A \to \Bbbk$ will make A into a Frobenius algebra over $\Bbbk$, for example, the quaternions $\mathbb{H}$ form a Frobenius algebra over $\mathbb{R}$. (Recall that $\mathbb{H} = \mathbb{R}1 \oplus \mathbb{R}i \oplus \mathbb{R}j \oplus \mathbb{R}k$ with multiplication defined by $i^2 = j^2 = -1$ and $ij = -ji = k$.)

2.2.16 Matrix algebras. The ring $\mathrm{Mat}_n(\Bbbk)$ of all n-by-n matrices over $\Bbbk$ is a Frobenius algebra with the usual trace map

$$\begin{aligned} \mathrm{Tr} : \mathrm{Mat}_n(\Bbbk) &\longrightarrow \Bbbk \\ (a_{ij}) &\longmapsto \sum_i a_{ii}. \end{aligned}$$

To see that the bilinear pairing resulting from Tr is nondegenerate, take the linear basis of $\mathrm{Mat}_n(\Bbbk)$ consisting of matrices E_{ij} with only one nonzero entry $e_{ij} = 1$; clearly E_{ji} is the dual basis element to E_{ij} under this pairing. Note that this is a symmetric Frobenius algebra (2.2.9) since the two matrix products

AB and BA have the same trace. If we twist the Frobenius form by multiplication with a noncentral invertible matrix we obtain a nonsymmetric Frobenius algebra.

As a concrete example, consider $\mathrm{Mat}_2(\mathbb{R}) = \{\left(\begin{smallmatrix} a & b \\ c & d \end{smallmatrix}\right) \mid a, b, c, d \in \mathbb{R}\}$ with the usual trace map

$$\begin{aligned} \mathrm{Tr} : \mathrm{Mat}_2(\mathbb{R}) &\longrightarrow \mathbb{R} \\ \left(\begin{smallmatrix} a & b \\ c & d \end{smallmatrix}\right) &\longmapsto a + d. \end{aligned}$$

Now twist and take as Frobenius form the composition

$$\begin{array}{ccccc} \mathrm{Mat}_2(\mathbb{R}) & \longrightarrow & \mathrm{Mat}_2(\mathbb{R}) & \xrightarrow{\mathrm{Tr}} & \mathbb{R} \\ \left(\begin{smallmatrix} a & b \\ c & d \end{smallmatrix}\right) & \longmapsto & \left(\begin{smallmatrix} a & b \\ c & d \end{smallmatrix}\right)\left(\begin{smallmatrix} 0 & 1 \\ 1 & 0 \end{smallmatrix}\right) & \longmapsto & b + c \end{array}$$

and check that this is not a central functional (i.e. does not satisfy the trace condition).

2.2.17 Semi-simple algebras of finite dimension. In fact all the above examples are special cases of semi-simple algebras. We will show in the Exercises that *every semi-simple algebra of finite dimension admits a Frobenius algebra structure* – in fact a symmetric one. To define semi-simplicity: the *Jacobson radical* $J(A)$ of a finite-dimensional $\mathbb{k}$-algebra A is the intersection of all left maximal ideals (or equivalently, the intersection of all right maximal ideals). Now a finite-dimensional $\mathbb{k}$-algebra A is *semi-simple* if its Jacobson radical is zero. There are several other characterisations, also for more general algebras and rings – see Curtis and Reiner [15] or Lang [30]. For example, the classical structure theorem of Wedderburn [49] states that *every finite-dimensional semi-simple algebra is isomorphic to a finite direct product of matrix algebras over skew-fields.*

2.2.18 Group algebras. (See for example Curtis and Reiner [15], Section 10.) Let $G = \{t_0, \ldots, t_n\}$ be a finite group written multiplicatively, and with $t_0 = 1$. The *group algebra* $\mathbb{k}G$ is defined as the set of formal linear combinations $\sum c_i\, t_i$ (where $c_i \in \mathbb{k}$) with multiplication given by the multiplication in G. It can be made into a Frobenius algebra by taking the Frobenius form to be the functional

$$\begin{aligned} \varepsilon : \mathbb{k}G &\longrightarrow \mathbb{k} \\ t_0 &\longmapsto 1 \\ t_i &\longmapsto 0 \quad \text{for } i \neq 0. \end{aligned}$$

Indeed, the corresponding pairing $g \otimes h \mapsto (gh)\varepsilon$ is nondegenerate since $g \otimes h \mapsto 1$ if and only if $h = g^{-1}$. It is also easy to see that this Frobenius algebra is symmetric. Again, we could get a nonsymmetric Frobenius algebra by twisting with some noncentral element (if there are any).

Concrete commutative example: let G be the group of nth roots of unity. Then the group algebra is isomorphic to $\Bbbk[x]/(x^n - 1)$. The Frobenius form defined as above takes $1 \mapsto 1$ and $x^i \mapsto 0$ for $i \neq 0 \mod n$.

Concrete noncommutative example: let $G = \mathfrak{S}_3$ be the symmetric group on three letters. It is generated by two transpositions, subject to three relations which make the group algebra look like this: $\Bbbk\langle x, y\rangle/(x^2 - 1, y^2 - 1, xyx - yxy)$. (Here the notation $\Bbbk\langle x, y\rangle$, with angle brackets, means noncommutative polynomial ring.) The Frobenius form sends $1 \mapsto 1$ and kills the subspace spanned by $\{x, y, xy, yx, xyx\}$.

If the characteristic of $\Bbbk$ does not divide the order of G then in fact $\Bbbk G$ is semi-simple (Maschke's theorem, cf. [15]), so this example is a subexample of the previous 2.2.17. In this case we can define a similar Frobenius form in a slick way like this: take ε to be the character of the regular representation. In detail, the *(first) regular representation* of G is just G acting on $\Bbbk G$ by right multiplication; we can regard this as a $\Bbbk$-algebra map $\Bbbk G \to \mathrm{End}(\Bbbk G)$. Its character is obtained by composition with the trace map (cf. 2.2.16). Since multiplication by $t_i \neq 1$ permutes all the t_j, we see that $1 \neq t_i \mapsto 0$, while clearly $t_0 = 1$ maps to the trace of the identity matrix, which is $n + 1$ (the order of the group). So this Frobenius form is nearly the same as the first one we constructed. We mention it here as a transition to:

2.2.19 Historical remarks. The *second regular representation* of a group G is

$$\begin{aligned}(\Bbbk G)^* \times G &\longrightarrow (\Bbbk G)^* \\ (\Lambda``, g) &\longmapsto [x \mapsto (gx)\Lambda],\end{aligned}$$

(compare 2.1.28). It is not difficult to show that the first and second regular representations of a finite group are isomorphic.

Now just as a representation of G can be viewed as a $\Bbbk$-algebra homomorphism $\Bbbk G \to \mathrm{End}(V)$, it makes good sense to speak about a representation of an algebra, say a finite-dimensional algebra A: it is a $\Bbbk$-algebra homomorphism $A \to \mathrm{End}(V)$. The first and second regular representations of A are $A \to \mathrm{End}(A)$ and $A \to \mathrm{End}(A^*)$, and it is a natural question to ask whether these two representations are isomorphic just as in the case of group algebras. The Prussian mathematician Ferdinand Georg Frobenius (1849–1917), founder of the theory of group representations, was the first to study algebras A with this property, which is equivalent to having $A \simeq A^*$ as right A-modules. This is the origin of the terminology 'Frobenius algebra'. According to the *Encyclopaedia of Mathematics* [26], Frobenius' first paper on the subject was *Theorie der hyperkomplexen Grössen* [22] from 1903. (*Hyperkomplexe Grössen*, or *hypercomplex numbers*, were that time's terminology for the elements in a finite-dimensional algebra, cf. also Wedderburn [49].)

This sort of algebra was given Frobenius' name in the late 1930s by T. Nakayama and C. Nesbitt in a series of papers in *Annals of Mathematics* about representations of algebras, e.g. Nakayama [41], *On Frobeniusean algebras I.* (The three (actually four) equivalent characterisations of Frobenius algebras listed in 2.2.7 date back to these papers, together with several others we have not touched upon.) In his 1938 article [42], Nesbitt writes:

> The writer, in collaboration with T. Nakayama, adopted the term Frobeniusean algebra, but now, quailing before our critics, we return to simply Frobenius algebra.

But Nakayama continued to use the term 'Frobeniusean' until 1950.

Note that in the classical theory, 'Frobenius algebra' just meant an algebra that admits a Frobenius structure, contrary to our usage where we require the structure to be specified.

2.2.20 The ring of group characters. (See Curtis and Reiner [15], Sections 30–31.) Assume the ground field is $\Bbbk = \mathbb{C}$. Let G be a finite group of order n. A *class function* on G is a function $G \to \mathbb{C}$ which is constant on each conjugacy class; the class functions form a ring denoted $R(G)$. In particular, the characters (traces of representations) are class functions, and in fact every class function is a linear combination of characters. There is a bilinear pairing on $R(G)$ defined by

$$\langle \phi \,|\, \psi \rangle := \frac{1}{n} \sum_{t \in G} \phi(t)\psi(t^{-1}).$$

Now the *orthogonality relations* (see [15], (31.8)) state that the characters form a orthonormal basis of $R(G)$ with respect to this bilinear pairing, so in particular the pairing is nondegenerate and provides a Frobenius algebra structure on $R(G)$.

2.2.21 Artinian Gorenstein rings. (See Eisenbud [21], Chapter 21.) Let A be a commutative Artinian local ring with maximal ideal $\mathfrak{m}$. The *socle* of A, denoted $\operatorname{Soc}(A)$, is the annihilator of $\mathfrak{m}$. The ring A is *Gorenstein* if $\operatorname{Soc}(A)$ is a *simple* A-module, meaning that there are no nontrivial submodules in $\mathrm{Soc}(A)$. Since A is a local ring this just means $\mathrm{Soc}(A) \simeq A/\mathfrak{m}$. Now we claim that if A is Gorenstein then $\mathrm{Soc}(A)$ is contained in every nonzero ideal of A. To establish this, we must show that $\mathrm{Soc}(A)$ lies inside the ideal (x) for every nonzero $x \in \mathfrak{m}$. Since $\mathrm{Soc}(A)$ is a 1-dimensional vector space (over $K := A/\mathfrak{m}$), it is enough to show that the two ideals intersect nontrivially. Now if x is already in $\mathrm{Soc}(A)$, then we are done. Otherwise there exists an element $y \in \mathfrak{m}$ such that xy is nonzero. But then (xy) is an ideal *strictly* smaller than (x) (by Nakayama's lemma, see [30]). Now repeat the argument with xy in place of x, and continue iteratively. Since A is Artinian, we cannot continue forever like that: eventually we arrive at a nonzero element in $\mathrm{Soc}(A)$, and we are done.

Now if A happens to be a finite-dimensional vector space over $\Bbbk$, then it follows that A can be made into a Frobenius algebra simply by taking any linear form which is nonzero on the socle. Indeed, since the nullspace of such a form does not contain the socle, it contains no nontrivial ideals at all.

In fact, conversely, every local Frobenius algebra is Gorenstein. We will not prove that here.

Some easy Gorenstein/Frobenius algebra examples: in $\Bbbk[x, y]/(x^2, y^2)$ the socle is generated by xy. In $\Bbbk[x]/(x^n)$ the socle is generated by x^{n-1}. (In general there is a number n such that $\mathfrak{m}^n = 0$ while $\mathfrak{m}^{n-1} \neq 0$. If A is Gorenstein then $\mathrm{Soc}(A) = \mathfrak{m}^{n-1}$.)

Here are two local rings which are not Gorenstein, and thus cannot support a Frobenius structure. In $A = \Bbbk[x, y]/(x^2, xy^2, y^3)$, the socle is (xy, y^2). In the ring $A = \Bbbk[x, y, z]/(x^2, y^2, z^2, xy)$, the socle is (xz, yz). The reason why there can be no Frobenius structure on these rings is the same in both cases (and does in fact work whenever $A/\mathfrak{m} \simeq \Bbbk$). Since the socle is of dimension at least 2, the nullspace of any linear form Λ will intersect $\mathrm{Soc}(A)$ nontrivially, so there exists a nonzero $s \in \mathrm{Soc}(A)$ with $s\Lambda = 0$. But this implies the whole ideal (s) is killed by Λ. Indeed, any element $a \in A$ can be written $a = u + m$ where $u \in \Bbbk$ and $m \in \mathfrak{m}$. So $as\Lambda = us\Lambda + ms\Lambda$. But $us\Lambda$ is zero by $\Bbbk$-linearity of Λ, and ms is zero since s is in the socle, so by definition it annihilates $\mathfrak{m}$.

2.2.22 Jacobian algebras. (See Griffiths and Harris [25], Chapter 5.1.) Let f be a polynomial in n variables, and suppose the zero locus $Z(f) \subset \mathbb{C}^n$ has an isolated singularity at $0 \in \mathbb{C}^n$. Put $f_i := \frac{\partial f}{\partial z_i}$ and let $I = (f_1, \ldots, f_n) \subset \mathcal{O}_0$ (the local ring at the origin). The local ring $\mathcal{O}_0/I$ is called a *Jacobian algebra*.

Since I is generated by n elements which is also its codimension, $\mathcal{O}_0/I$ is a complete intersection ring and in particular Gorenstein. But more interestingly, there is a canonical Frobenius form on it, defined by integrating around the singularity along a real n-ball. Precisely, let $B = \{z \mid f_i(z) = \rho\}$ (for some small $\rho > 0$), and let the functional be the residue

$$\begin{aligned} \mathrm{res}_f : \mathcal{O}_0/I &\longrightarrow \mathbb{C} \\ g &\longmapsto \left(\tfrac{1}{2\pi i}\right)^{2n} \int_B \frac{g(z) \cdot dz_1 \wedge \cdots \wedge dz_n}{f_1(z) \cdots f_n(z)}. \end{aligned}$$

Now local duality (see Griffiths and Harris [25], page 659) states that the corresponding bilinear pairing is nondegenerate.

2.2.23 Cohomology rings. (See for example Bott and Tu [12], Chapter 1, or Fulton [23], 24.32.) To be concrete, let X be a compact oriented manifold of dimension n, and let $H^*(X) = \oplus_{i=0}^n H^i(X)$ denote the de Rham cohomology ($H^i(X) =$ closed differentiable i-forms modulo the exact ones). It is a ring under the wedge product. Integration over X (with respect to a chosen volume form) provides a linear map $H^*(X) \to \mathbb{R}$, and Poincaré duality states that the corresponding bilinear pairing $H^*(X) \otimes H^*(X) \to \mathbb{R}$ is nondegenerate; precisely, $H^i(X)$ is dual to $H^{n-i}(X)$. Thus, $H^*(X)$ is a Frobenius algebra over $\mathbb{R}$.

In fact, if X is connected then $H^*(X)$ is a (graded-commutative) Gorenstein ring (2.2.21): the maximal ideal is $\bigoplus_{i>0} H^i(X)$, and the socle is $H^n(X) \simeq \mathbb{R}$. By *graded-commutative* we mean that classes of odd degree anti-commute: given $\alpha \in H^p(X)$ and $\beta \in H^q(X)$ then

$$\alpha \wedge \beta = (-1)^{pq} \beta \wedge \alpha.$$

As a concrete example, take the cohomology ring of a torus X: it is generated by two 1-forms α and β (Poincaré dual to the two nonhomologous circles generating $H_1(X)$ ('one around the hole and one through the hole')). Then we have

$$H^*(X) = \mathbb{R}\langle \alpha, \beta \rangle / (\alpha^2 = \beta^2 = 0, \alpha\beta = -\beta\alpha).$$

Another basic example is the cohomology of $X = \mathbb{CP}^n$. It is $H^*(X) \simeq \mathbb{R}[h]/h^{n+1}$, where h is the class of a hyperplane.

Exercises

1. Prove that the four characterisations of symmetric Frobenius algebras (2.2.9) are equivalent.
2. Prove Lemma 2.2.11.
3. Show that every group algebra admits a symmetric Frobenius algebra structure.
4. Combine 2.2.16 and 2.2.13 to show that a matrix ring over $\mathbb{C}$ is also a Frobenius algebra over $\mathbb{R}$. Can you also prove that a matrix ring over $\mathbb{H}$ is a Frobenius algebra over $\mathbb{R}$?

$\star$5. More generally, show that *a matrix algebra over a Frobenius algebra is again a Frobenius algebra.* Precisely, let (A, ε) be a Frobenius algebra over $\Bbbk$, and let $M_n(A)$ denote the algebra of n-by-n matrices over A, with the usual trace map $\mathrm{Tr} : M_n(A) \to A$, $(m_{ij}) \mapsto \sum_i m_{ii}$. Show that the composite

$$M_n(A) \xrightarrow{\mathrm{Tr}} A \xrightarrow{\varepsilon} \Bbbk$$

is a Frobenius form. (Hint: assume the dimension of A over $\Bbbk$ is r, then $M_n(A)$ has dimension rn^2. Pick a basis for A and consider the basis for $M_n(A)$ consisting in matrices with a single nonzero entry which belongs to the basis of A.)
Since every finite-dimensional simple algebra is a matrix algebra over a skew-field, we have shown that *every finite-dimensional simple algebra admits a Frobenius algebra structure.*

6. Show that every finite-dimensional simple algebra admits a symmetric Frobenius structure.
7. Given two Frobenius algebras (A', ε') and (A'', ε''), show that the linear functional

$$\begin{aligned} A' \oplus A'' &\longrightarrow \Bbbk \\ a' \oplus a'' &\longmapsto a'\varepsilon' + a''\varepsilon'' \end{aligned}$$

is a Frobenius form on the direct product algebra $A' \times A'' = A' \oplus A''$ (cf. Exercise 8 on page 93).
Conversely, show that if A is a direct product algebra $A = A' \times A''$, and if $\varepsilon : A \to \Bbbk$ is a Frobenius form on A, then the composites $A' \to A \to \Bbbk$ and $A'' \to A \to \Bbbk$ are Frobenius forms on A' and A'' respectively. (Note that $A' \to A$ and $A'' \to A$ are just $\Bbbk$-linear maps, not ring homomorphisms.)

8. The previous exercise shows that Frobenius structure is compatible with direct products. Redo that exercise using the definition of Frobenius algebra in terms of a Frobenius pairing.
9. Use Wedderburn's structure theorem (cf. 2.2.17) and the previous exercises to prove that *every semi-simple algebra of finite dimension admits a Frobenius structure*.
10. Show that every finite-dimensional semi-simple algebra admits a symmetric Frobenius structure.
11. (A noncommutative analogue of 2.2.21.) Let A be a finite-dimensional $\Bbbk$-algebra; let J denote its Jacobson radical, and let S be the right annihilator of J:

$$S := \{a \in A \mid xa = 0\ \forall x \in J\}.$$

This is a two-sided ideal. Assume that S is a simple left ideal, i.e. contains no nontrivial left ideals. Show that S is contained in every left ideal, and conclude that a Frobenius form on A can be taken to be any linear functional which is nonzero on S. All the arguments are in 2.2.21 – you just need to refine them to the noncommutative case...
12. (Nonexample.) Show that the algebra of upper-triangular 2-by-2 matrices over $\Bbbk$ does not admit a Frobenius structure.
13. (Nakayama.) Let (A, ε) be a Frobenius algebra with pairing β given by $\langle x \mid y \rangle := (xy)\varepsilon$. Consider the map $\sigma : A \to A$ defined as the composite

$$A \xrightarrow{\beta_{\text{right}}} A^* \xrightarrow{\beta_{\text{left}}^{-1}} A,$$

cf. 2.2.5 for notation.
(i) Show that σ is characterised by the formula $\langle x \mid y \rangle = \langle y \mid x\sigma \rangle$, $\forall x, y \in A$. (In particular, (A, ε) is symmetric if and only if σ is the identity map.)
(ii) Show that σ is a $\Bbbk$-algebra homomorphism.

2.3 Frobenius algebras and comultiplication

2.3.1 Coalgebras. The notion of coalgebra over $\Bbbk$ is the opposite of the notion of $\Bbbk$-algebra, in the sense that the structure maps and the diagrams for their axioms are all just reversed. So a *coalgebra* over $\Bbbk$ is a vector space A together with two $\Bbbk$-linear maps

$$\delta : A \to A \otimes A, \qquad \varepsilon : A \to \Bbbk$$

such that these three diagrams commute:

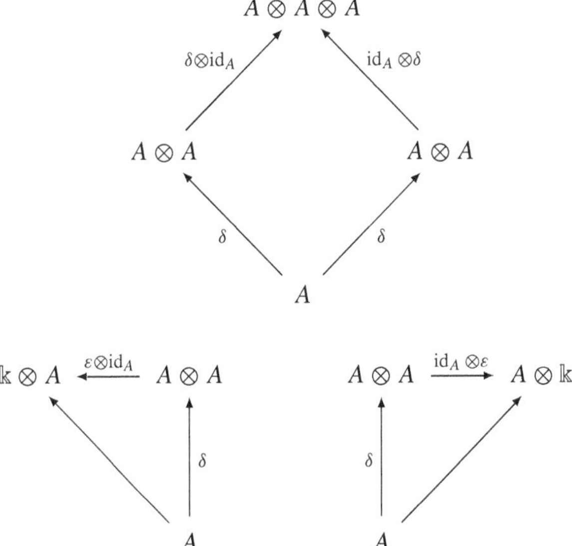

(Compare 2.1.18.) The map δ is called *comultiplication*, and $\varepsilon : A \to \Bbbk$ is called the *counit* (or sometimes the *augmentation*). The axioms expressed in the diagrams are called *coassociativity* and the *counit condition*.

Here we are happy to express things in terms of diagrams, because it is very messy to write down anything explicitly about comultiplication. To do that we would have to use coordinates. We will briefly take that viewpoint in the subsection starting on page 123.

2.3.2 Example: the coalgebra on a set. Let $S = \{t_0, \ldots, t_n\}$ be a set. Then the *coalgebra on* S is given by taking the vector space spanned by S, with comultiplication given by the diagonal map $S \to S \times S$. Precisely, let $V := \Bbbk S$ be the set of formal linear combinations of the elements in S. The comultiplication is given by letting $t_i \mapsto t_i \otimes t_i \in V \otimes V$ and extending linearly to the whole of V. The counit is given by $t_i \mapsto 1$ for each t_i. (If S has group structure then $\Bbbk S$ acquires algebra structure, and we recover the definition of group algebra, cf. 2.2.18.)

A similar way of saying essentially the same: as soon as a basis has been given for a vector space V it acquires a canonical coalgebra structure.

2.3.3 Example: the trigonometric coalgebra of Sweedler [46]. Let V be a 2-dimensional vector space with basis $\{S, C\}$. Define a comultiplication by

$$\begin{aligned} C &\mapsto C \otimes C - S \otimes S \\ S &\mapsto C \otimes S + S \otimes C, \end{aligned}$$

with counit given by

$$\begin{aligned} C &\mapsto 1 \\ S &\mapsto 0, \end{aligned}$$

and check that coassociativity and the counit condition hold. The reason for the name is the analogy with the familiar formulae

$$\begin{aligned} \cos(x+y) &= \cos(x)\cos(y) - \sin(x)\sin(y), \\ \sin(x+y) &= \cos(x)\sin(y) + \sin(x)\cos(y), \\ \cos(0) &= 1, \\ \sin(0) &= 0. \end{aligned}$$

2.3.4 Towards coalgebra structure on a Frobenius algebra. It is not a coincidence that we have denoted the counit by ε just like the Frobenius form. The main result of this chapter states that *every Frobenius algebra has a unique coalgebra structure for which the Frobenius form is the counit, and which is A-linear* (see Proposition 2.3.22 for the exact statement). And conversely, *given a $\Bbbk$-algebra, equipped with an A-linear coalgebra structure, then the counit is a Frobenius form* (cf. Proposition 2.3.24). So this gives yet another characterisation of Frobenius algebras – the most important one for our purposes. We will give a quite elementary proof, which does not even involve coordinates. It is based on a graphical calculus which is common in knot theory and quantum groups (see Kassel [29]). Usually the pictures are graphs of various sorts. Here, inspired by the pictures in Chapter 1 we adopt a graphical representation which looks like topological surfaces with boundary, and in the end the notion of TQFT is going to give a precise interpretation to that analogy, cf. 3.3.2.

Graphical calculus

2.3.5 The building blocks. The first observation is that we do not have many pieces to move! If we want to construct a comultiplication on our Frobenius algebra A, all we have to make do with are the following maps: the multiplication $\mu : A \otimes A \to A$, the unit $\eta : \Bbbk \to A$, and the Frobenius form $\varepsilon : A \to \Bbbk$ as well as the Frobenius pairing $\beta : A \otimes A \to \Bbbk$, not forgetting the identity map $\mathrm{id}_A : A \to A$. These maps come with certain properties which are expressed as commutative diagrams. Our task is to combine these arrows in a natural way to construct a comultiplication, and then combine all the diagrams in order to establish the diagrams that express the properties we want from this comultiplication.

2.3.6 Towards graphical representation of the structures. The second observation is that *all of these building blocks are maps between tensor powers of* A; let A^n denote the tensor product of n copies of A. Of course the ground field appears in the maps, but recall that it is natural to consider $\Bbbk$ as the zeroth tensor power of A, the tensor product with zero factors.

In my very first mathematics book, in the first year of primary school, there were a lot of drawings meant to make the learning of multiplication funnier and more conceptual. Multiplication was introduced as a machine with two input holes, where you could throw in two numbers, and then the machine would process this input and produce a number which would drop out from a single output hole. The machines looked like this:

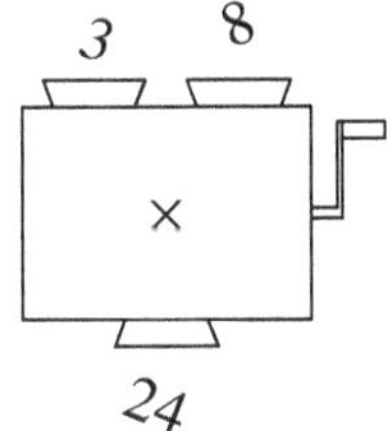

We will basically adopt this picture, but we will turn it 90 degrees and draw the multiplication map like this:

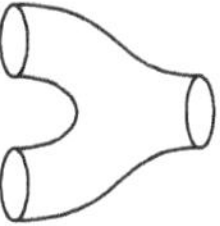

2.3.7 The dictionary (or pictionary). Let us first draw the maps that define a $\Bbbk$-algebra:

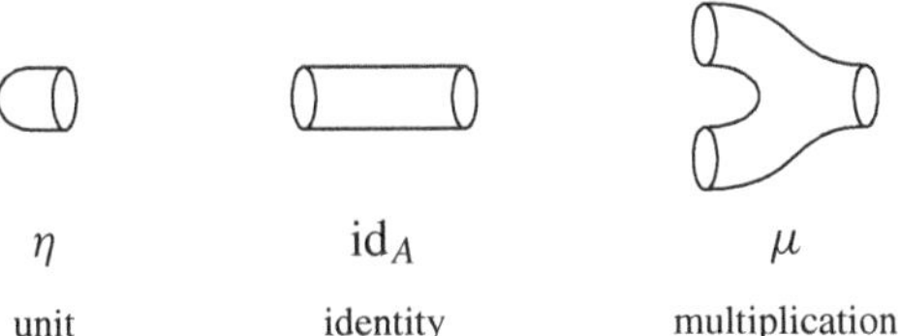

η	id_A	μ
unit	identity	multiplication

(The identity map is actually not a part of the algebra structure – it is something even more fundamental, since it is already part of the vector space structure on A (in fact it is automatically present on any object in any category…).)

We will now give precise meaning to each picture: throughout this chapter, these symbols have the status of formal mathematical symbols, just like the symbols $\rightarrow$ or $\otimes$. The symbol corresponding to each $\Bbbk$-linear map $\phi : A^m \rightarrow A^n$ has m boundaries on the left (input holes): one for each factor of A in the source, and ordered such that the first factor in the tensor product corresponds to the bottom input hole and the last factor corresponds to the top input hole. If $m = 0$ we simply draw no in-boundary. Similarly there are n boundaries on the right (output holes) which correspond to the target A^n, with the same convention for the ordering.

The tensor product of two maps is drawn as the (disjoint) union of the two symbols – one placed above the other, in accordance with our convention for ordering. Indeed, the tensor product of two maps is defined by letting the two maps operate independently on their respective arguments, so it is natural that we draw this as two parallel tubes or, in the machine metaphor, as two parallel processes – and similarly for multiple tensor products.

So for example the map $\mathrm{id}_A \otimes \mu$ has symbol

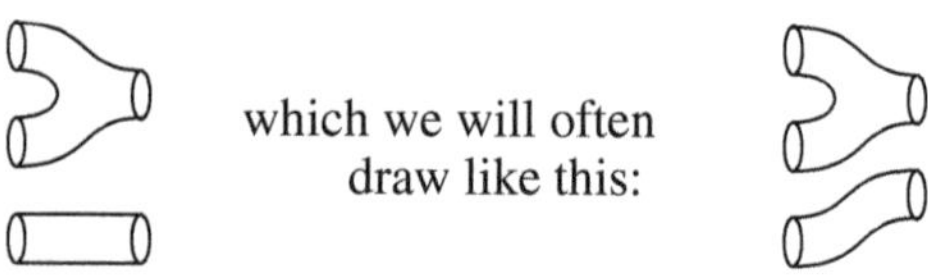

just for graphical convenience.

It is clear how the graphical language should capture compositions: just join the output holes of the first figure with the input holes of the second. For example, the composition

$$\Bbbk \otimes A \xrightarrow{\eta \otimes \mathrm{id}_A} A \otimes A \xrightarrow{\quad\mu\quad} A$$

is represented by

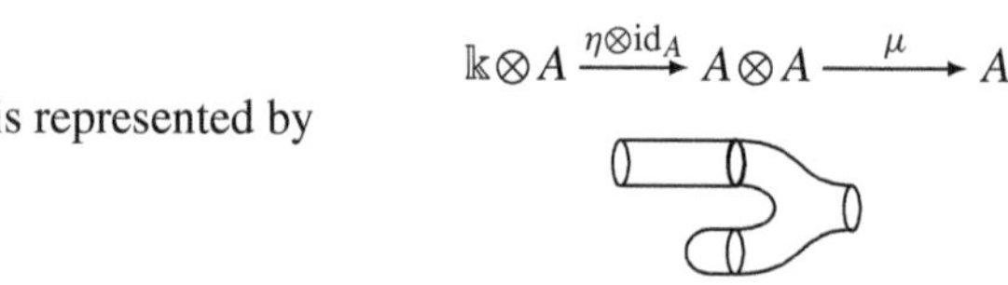

Now we can write down the axioms for an algebra in graphical notation. The 'algebra-book' version of the axioms (like $(ab)c = a(bc)$) cannot easily be expressed in graphical terms because we have no good way to treat elements, but the way we have expressed the axioms in 2.1.18 in terms of diagrams, we are talking about the equality of two different compositions of arrows. It is easy to express 2.1.18 in graphical terms:

2.3.8 The $\Bbbk$-algebra axioms.

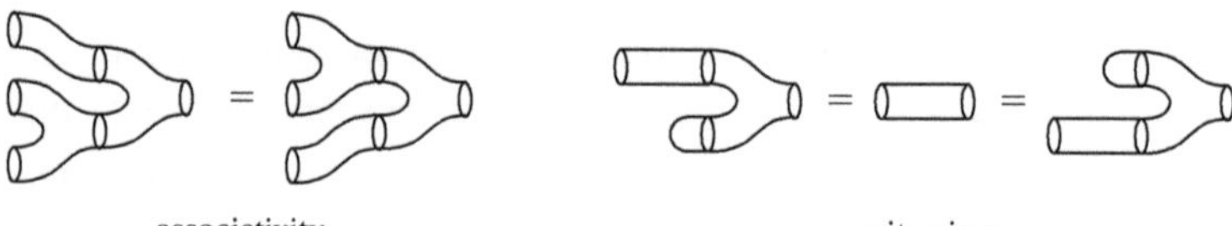

It should be stressed that this is not just an analogy or a fancy illustration. The symbols above have precise mathematical meaning and express exactly the same conditions as the diagrams in 2.1.18.

2.3.9 Frobenius form and Frobenius pairing. We want to express Frobenius structure in graphical language. We have three equivalent definitions, but the last one (2.2.6), the A-isomorphisms $A \xrightarrow{\sim} A^*$, does not fit into our graphical notation because A^* is not a tensor power of A.

The first definition (2.2.1) involves a linear form $\varepsilon : A \to \Bbbk$, and the second a bilinear pairing $\beta : A \otimes A \to \Bbbk$. According to our principles we depict those two maps as

Frobenius form Frobenius pairing

We can draw right away the relation between these two maps:

These are the relations $\langle x \,|\, y \rangle = (xy)\varepsilon$ and $\langle 1_A \,|\, x \rangle = x\varepsilon = \langle x \,|\, 1_A \rangle$ explained in 2.2.3. Here come the diagrams:

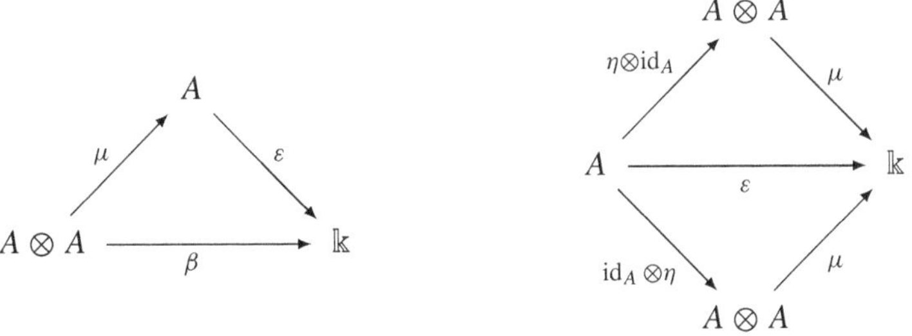

Note that we suppress the $\Bbbk$-factors in tensor products, writing $A \xrightarrow{\mathrm{id}_A \otimes \eta} A \otimes A$ when we really mean $A \otimes \Bbbk \xrightarrow{\mathrm{id}_A \otimes \eta} A \otimes A$.

It is trickier to express the axioms which ε and β must satisfy in order to be a Frobenius form and a Frobenius pairing, respectively. The axiom for a Frobenius form $\varepsilon : A \to \Bbbk$ is that its nullspace contains no nonzero ideals. We cannot express this condition in graphical language because we have no way to represent an ideal... In contrast, the axiom for the Frobenius pairing allows graphical expression. There are two conditions: associativity (2.1.32) and nondegeneracy (2.1.10).

2.3.10 Associativity of the Frobenius pairing. The graphical expression of the associativity condition on β is

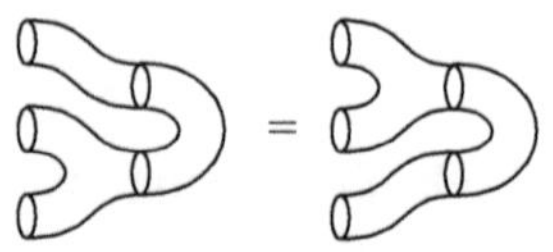

associativity
of the pairing β

In view of the relation = , the associativity equation for β follows from the associativity equation for μ, simply by putting a cap on the output hole of the drawing of 2.3.8.

2.3.11 Nondegeneracy – the snake relation. According to the definition (2.1.10), nondegeneracy means the *existence of a copairing* $\gamma : \Bbbk \to A \otimes A$ *such that these two diagrams commute:*

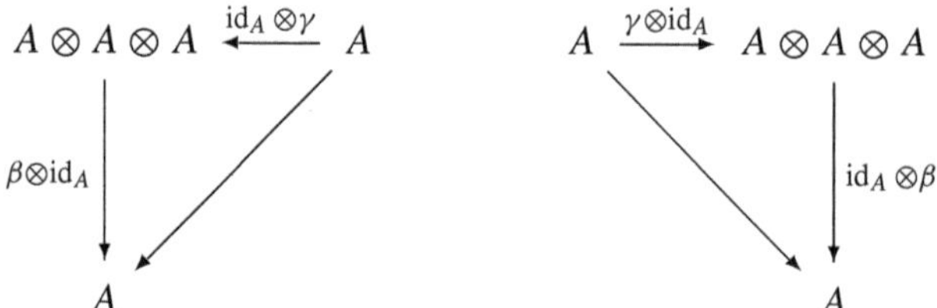

Now we are in business, because these are diagrams of maps between tensor powers of A. Here is the graphical expression:

There exists *such that*

= =

This is really the crucial property – we will henceforth refer to this as *the snake relation.*

Our goal is to show that a Frobenius algebra (A, ε) has a natural coalgebra structure for which ε is the counit. To construct a comultiplication on A, the key element is exactly the copairing , since it serves to turn around an input hole so that it becomes an output hole. We will simply put

:= =

Our first task is to show that these two definitions coincide (which is the content of Lemma 2.3.15). To this end, it is practical to introduce the *three-point function* (it is usually called this in field theory).

2.3.12 The three-point function $\phi : A \otimes A \otimes A \to \Bbbk$ is defined by

$$\phi := (\mu \otimes \mathrm{id}_A)\beta = (\mathrm{id}_A \otimes \mu)\beta,$$

which in graphical language reads

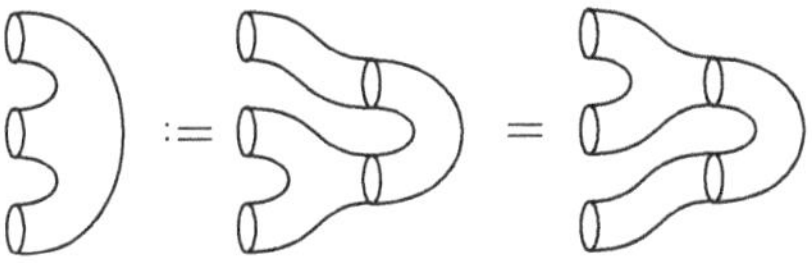

Associativity of β says that the two expressions coincide. (In other words, $\phi(a, b, c) = \langle ab \,|\, c \rangle = \langle a \,|\, bc \rangle$.) More figuratively, we can say that the pairing is used to turn around the output hole of  ; then associativity states that it does not matter which way we turn around.

Conversely, using the snake relation we can express in terms of the three-point function:

2.3.13 Lemma. *We have*

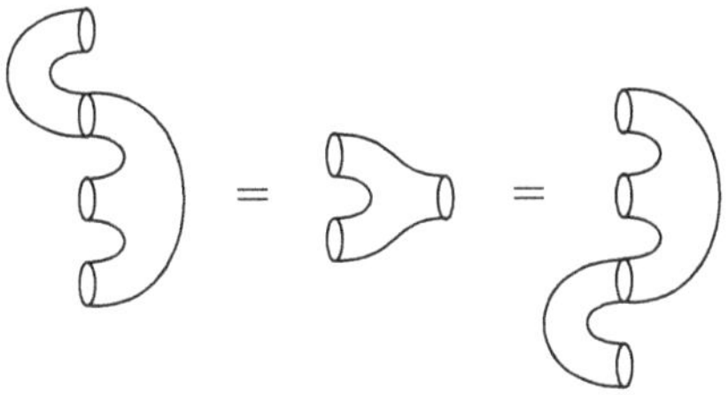

Proof. This is our first graphical proof, so let us walk through it slowly. We concentrate on the left-hand equation – the right-hand equation is completely analogous. First of all we should explain what is meant by the drawing. As it stands it does not really represent a composition – the input and output holes do not match! We have omitted some identity maps. What we mean is really

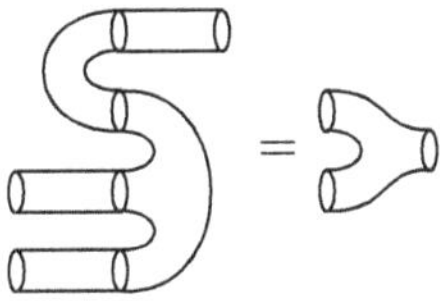

This omission is harmless, so for simplicity we will most often write like this. Here in this proof however, we will write out everything – in order to illustrate the harmlessness of the omission.

First let us write down the statement in terms of a diagram:

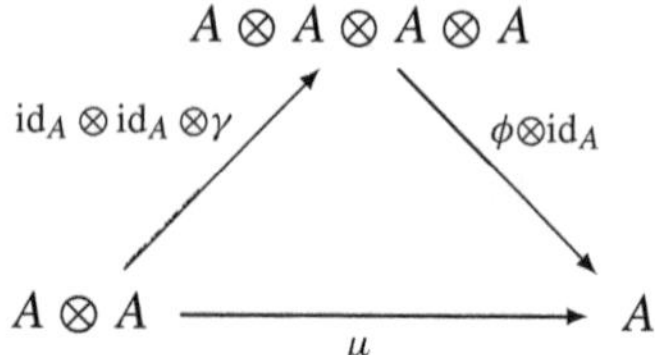

Note that we omit the $\Bbbk$-factors.

Now for the proof. First use the definition of the three-point function:

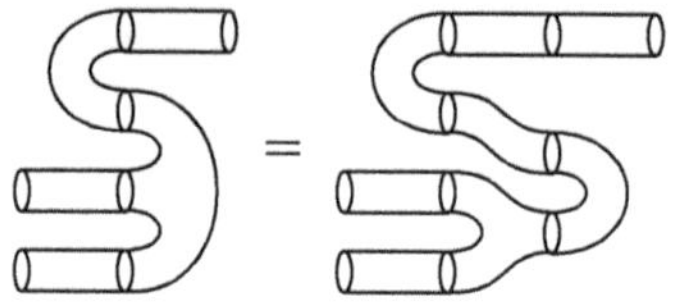

Now remove four identity maps, and insert a new one just after the multiplication (in order to line up things to our advantage):

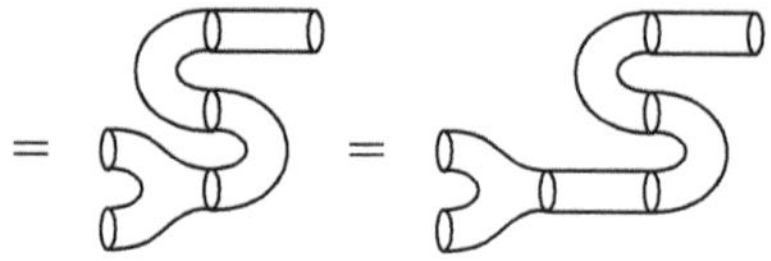

Now use the snake relation – this is the crucial step – and finally remove an identity map:

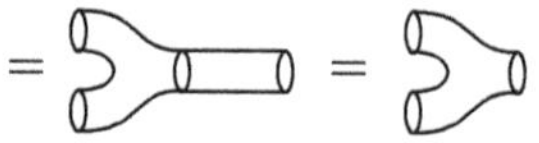

□

To stress that each single step in the graphical proof is in fact a commutative diagram, let us rephrase the whole proof!

2.3.14 Diagram proof of 2.3.13. We started with this composition

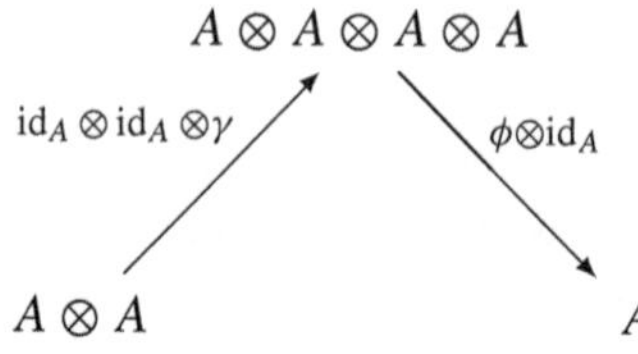

and the first step was to use the diagram expressing the definition of ϕ, to fill in this triangle:

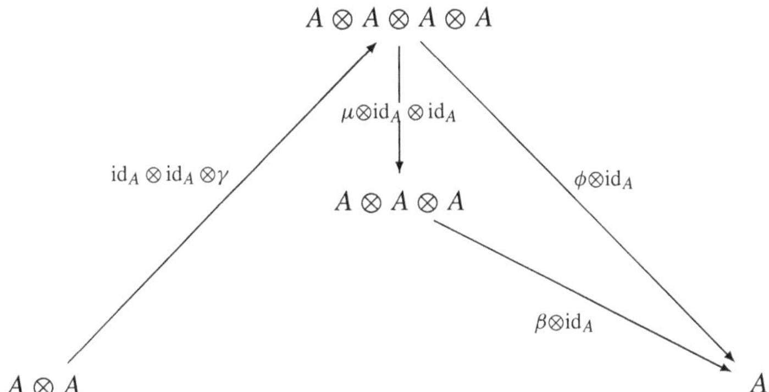

The second step was to notice that those identity maps in the first two arrows were superfluous – that amounts to filling in another triangle:

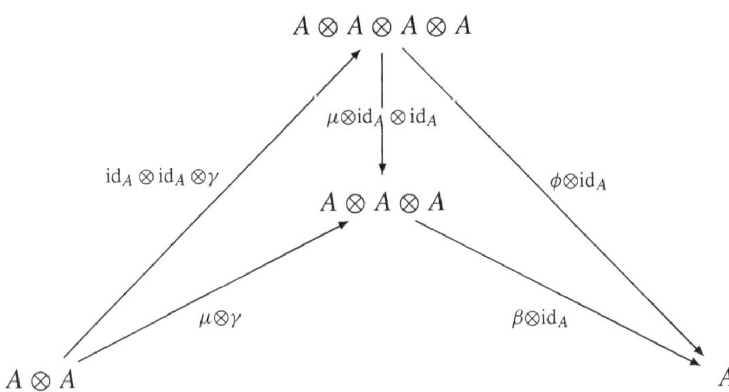

Next, we inserted a new identity map like this:

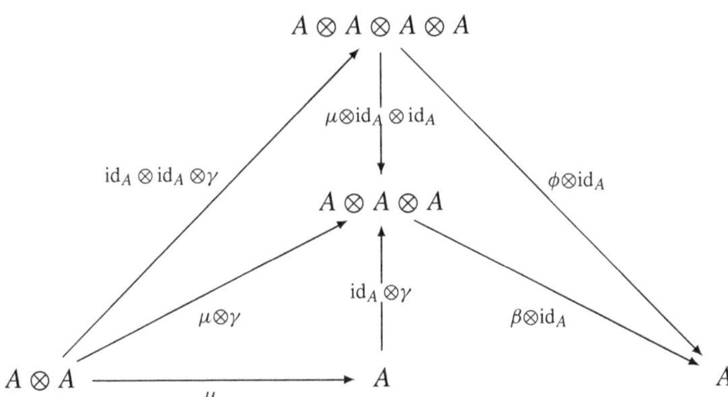

Finally, we used the fact that β is nondegenerate (diagram 2.1.10), to fill in the last triangle with an identity map id_A, and noticed that the composite $\mu\,\mathrm{id}_A$ is just μ itself (the curved arrow):

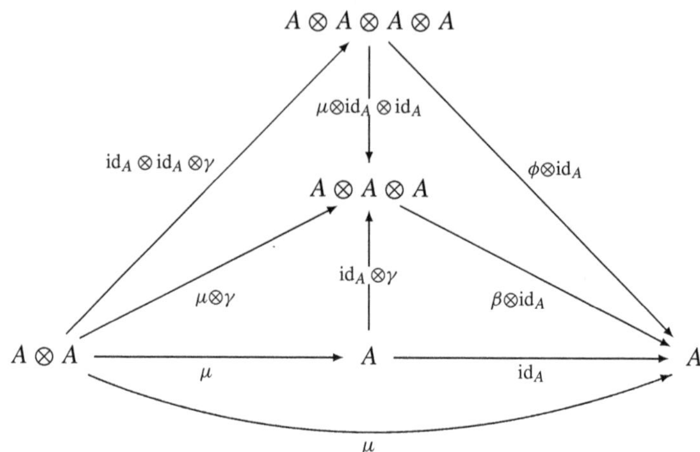

which is what we wanted to prove. □

2.3.15 Lemma. *We have*

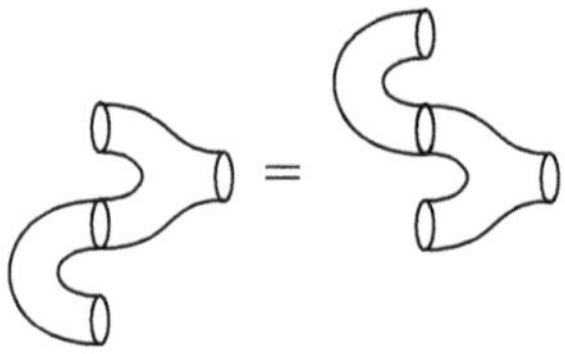

Proof. Follows immediately using the expressions for given in 2.3.13 □

2.3.16 Comultiplication. Now we define a comultiplication δ

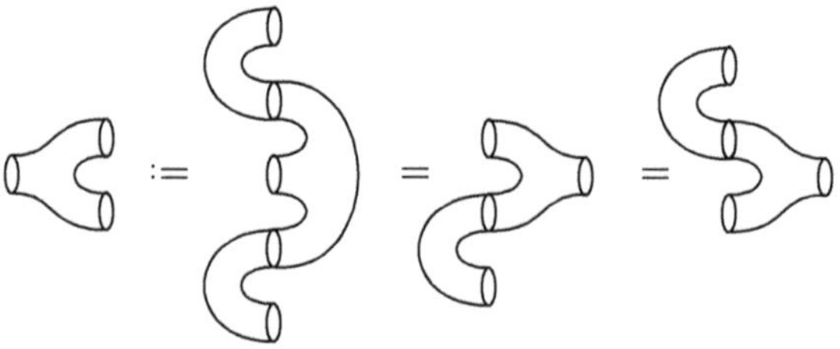

This makes sense, due to 2.3.13

2.3.17 Multiplication in terms of comultiplication. Conversely, turning some holes back again, using β, and then using the snake relation, we also get the relations dual to 2.3.16:

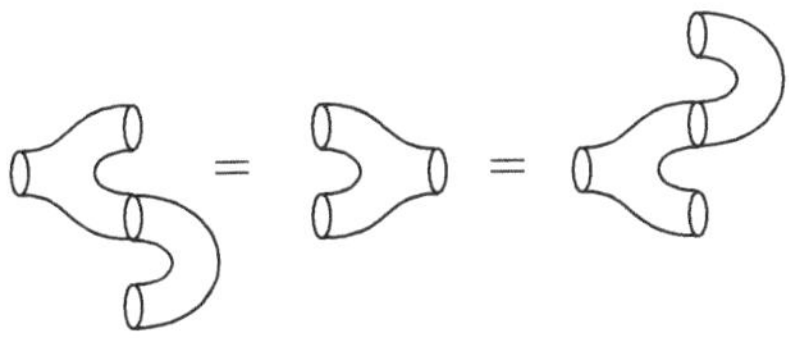

2.3.18 Lemma. *The Frobenius form ε is counit for δ:*

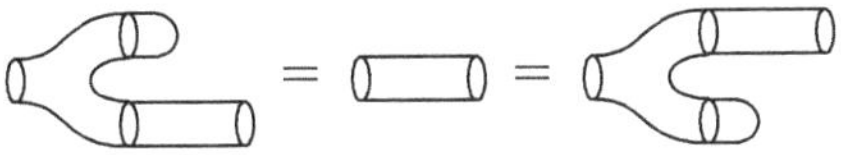

Proof. Suppressing the identity maps, write

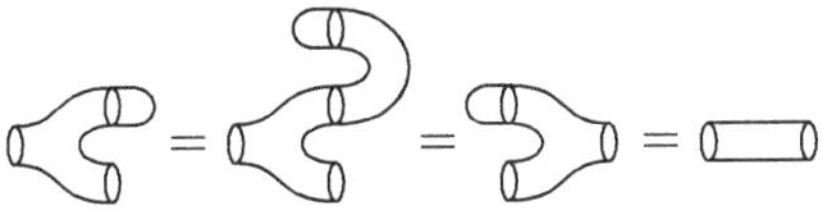

Here the first step was to use the expression 2.3.9 for . The next step was to use relation 2.3.17. Finally we used that is neutral element for the multiplication (cf. 2.3.8). (The right-hand equation is analogous.) □

2.3.19 Lemma. *The comultiplication δ defined above satisfies the following relation, called the* Frobenius condition.

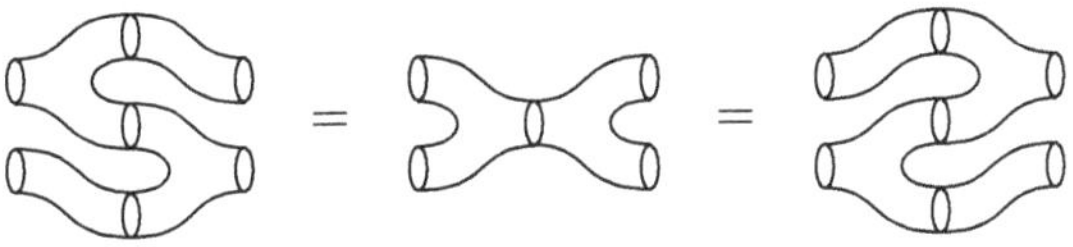

The right-hand equation amounts to the commutativity of this diagram:

$$\begin{array}{ccc} A \otimes \underline{A} & \xrightarrow{\delta \otimes \mathrm{id}_A} & A \otimes A \otimes \underline{A} \\ \mu \downarrow & & \downarrow \mathrm{id}_A \otimes \mu \\ A & \xrightarrow[\delta]{} & A \otimes A \end{array}$$

which in turn expresses right A-linearity of δ. (The underlined copies of A are those that act by scalar multiplication.) Similarly the left-hand equation expresses left A-linearity.

Proof. For the left-hand equation, use = ; then use associativity, and finally use the relation back again:

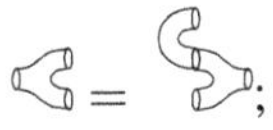

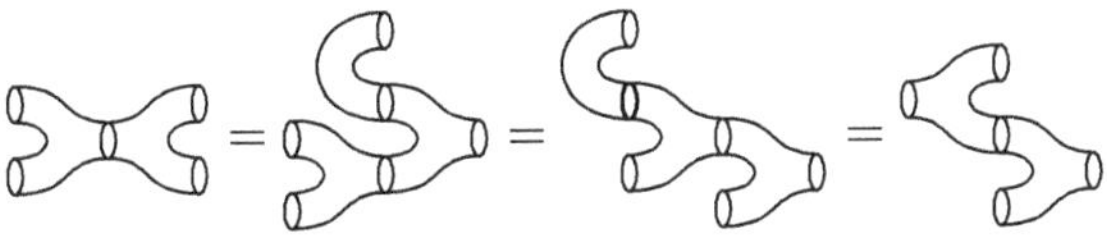

The right-hand equation is obtained using 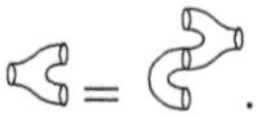 . □

2.3.20 Lemma. *The comultiplication is coassociative:*

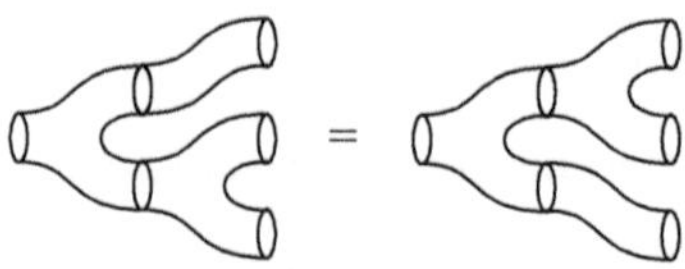

Proof. Use the definition of δ (2.3.16), then the associativity, and finally the definition again:

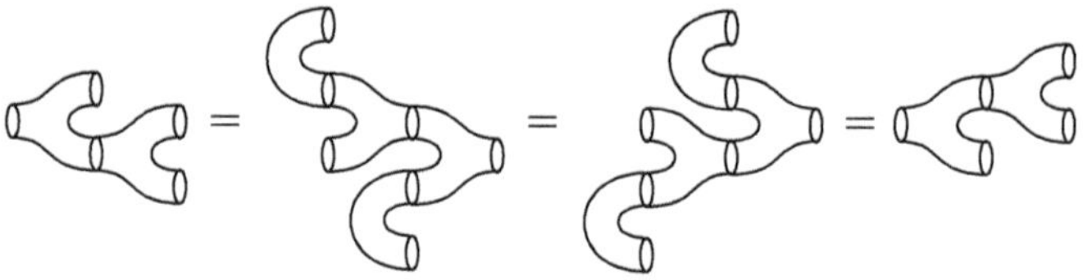

□

The next lemma shows that the relation between the copairing and the unit is analogous (dual) to the relation between the Frobenius form and the Frobenius pairing (relation 2.3.9). (Its proof is left as an exercise.)

2.3.21 Lemma. *These relations hold:*

□

2.3.22 Proposition. *Given a Frobenius algebra (A, ε), there exists a unique comultiplication whose counit is ε and which satisfies the Frobenius relation, and this comultiplication is coassociative.*

Proof. We have already constructed such a comultiplication (2.3.16, 2.3.18, 2.3.19), and established its coassociativity (2.3.20). It remains to show that it is unique. This is a consequence of the fact that the copairing corresponding to a nondegenerate pairing is unique, as we proved in 2.1.11. Let us repeat that proof in graphical language – anyway we will need this version later on. Let $\beta =$ be a pairing and suppose that ξ and γ are two corresponding copairings:

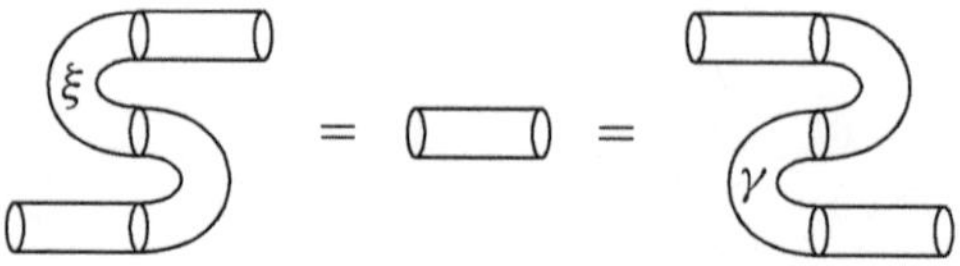

Then we get

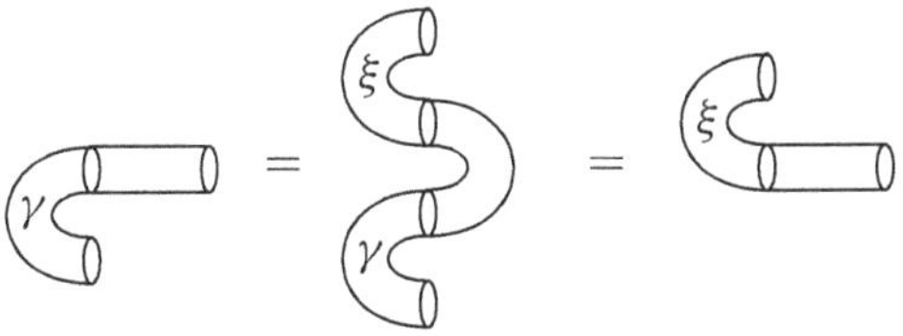

(2.3.23)

which is to say that ξ and γ coincide.

Now for the uniqueness of the comultiplication: suppose that ω is another comultiplication with counit ε and which satisfies the Frobenius relation. In the following couple of arguments we work with the left-hand side of the Frobenius relation for ω, establishing half of the relations we need. The other relations follow by applying similar arguments to the right-hand side. Putting caps on the upper input hole and the lower output hole of the Frobenius relation we see that $\eta\omega$ satisfies the snake equation:

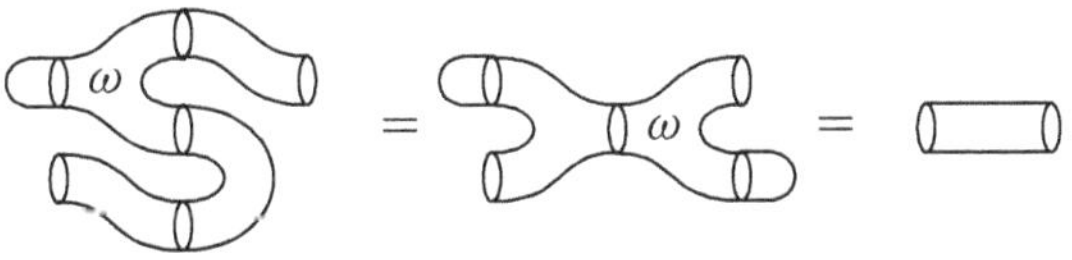

by the unit and counit axioms. So by the uniqueness of copairing we have $\eta\omega = \gamma$. Using this, if instead we put only the cap η on, then we get

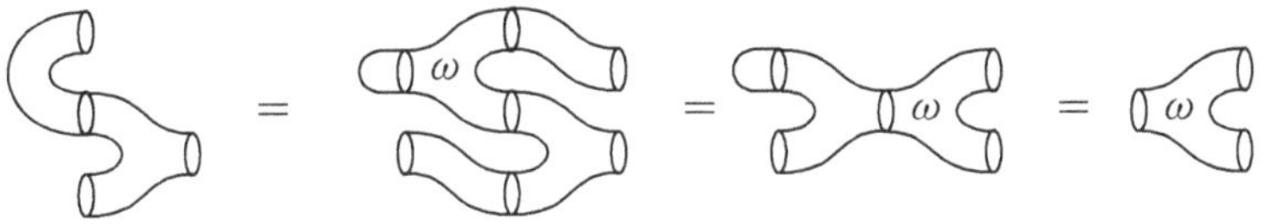

That is, ω is nothing but μ with an input hole turned around, just like δ was defined. □

The reason why the relation of 2.3.19 is called the Frobenius condition is that it characterises Frobenius algebras, as the next result shows. In fact, it characterises Frobenius algebras not only among the associative algebras of finite dimension, but also among general vector spaces equipped with unitary multiplication. Precisely,

2.3.24 Proposition. *Let A denote a vector space equipped with a multiplication map $\mu : A \otimes A \to A$ denoted , with unit $\eta : \Bbbk \to A$ denoted , a comultiplication $\delta : A \to A \otimes A$ denoted , with counit $\varepsilon : A \to \Bbbk$ denoted , and suppose the Frobenius relation holds:*

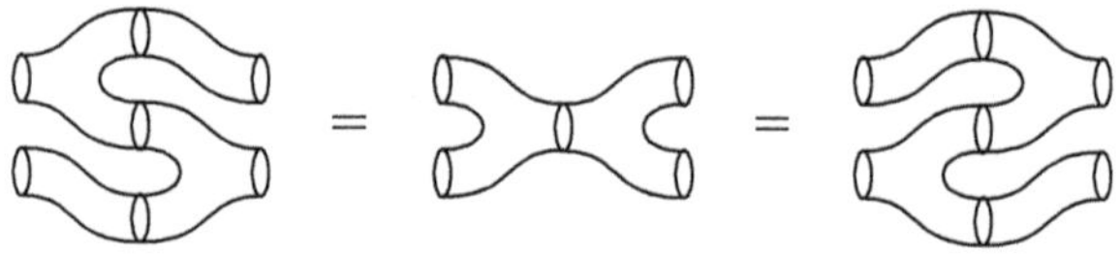

Then

(i) *the vector space A is of finite dimension,*
(ii) *the multiplication μ is associative, and thus A is a finite-dimensional $\Bbbk$-algebra (also, the comultiplication is coassociative),*
(iii) *the counit ε is a Frobenius form, and thus (A, ε) is a Frobenius algebra.*

Proof. Set $\beta := \mu\varepsilon$, that is: $=$ 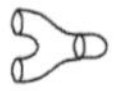 . We will show that β is nondegenerate, i.e. establish the snake relation, with $\gamma = \eta\delta$. The proof of this goes just like the previous proof. Put caps on the left-hand part of the Frobenius relation like this:

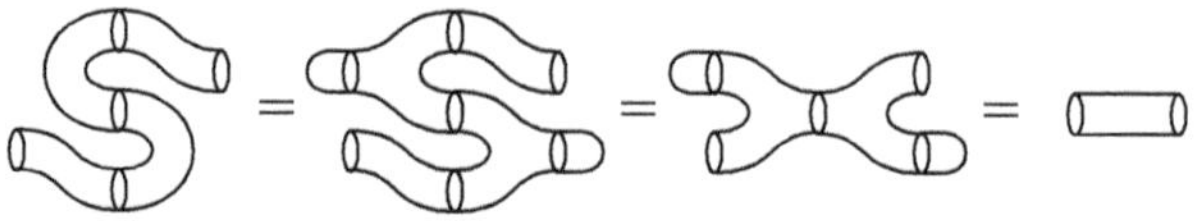

using the unit and counit axioms. This is the left-hand part of the snake relation; similarly, the right-hand side of the Frobenius relation gives the right-hand side of the snake relation, so β is nondegenerate. This in particular implies that A is of finite dimension (cf. 2.1.12).

To get associativity, put only one cap on the Frobenius relation (left-hand relation), getting these two identities:

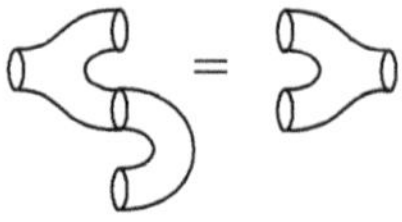

and

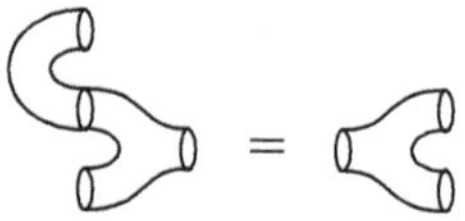

Now we can write

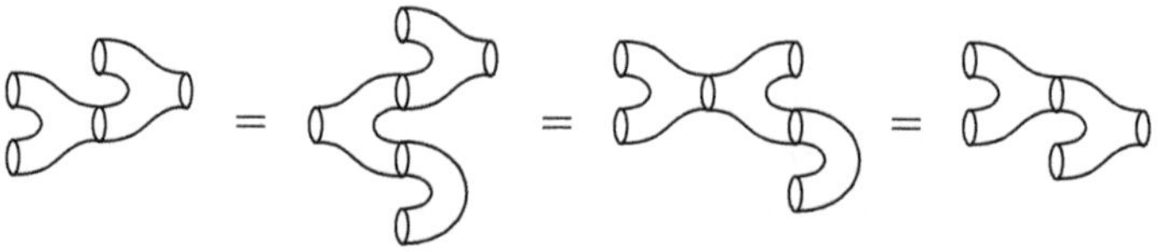

So is associative. (Coassociativity follows similarly.)

Finally, since is associative, clearly the pairing $=$ 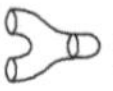 is associative as well, so (A, β) is a Frobenius algebra. □

2.3.25 Historical remarks. The characterisation of Frobenius algebras in terms of comultiplication (2.3.24) goes back (at least) to Lawvere [32] (1967), where it is a parenthetical remark at the end of the paper. In a very general categorical context (which we will take up in Chapter 3, notably 3.6.8) he describes a *Frobenius standard construction* (standard construction meaning monad) as being a combined monoid/comonoid object with this compatibility requirement (which we reproduce in graphical language):

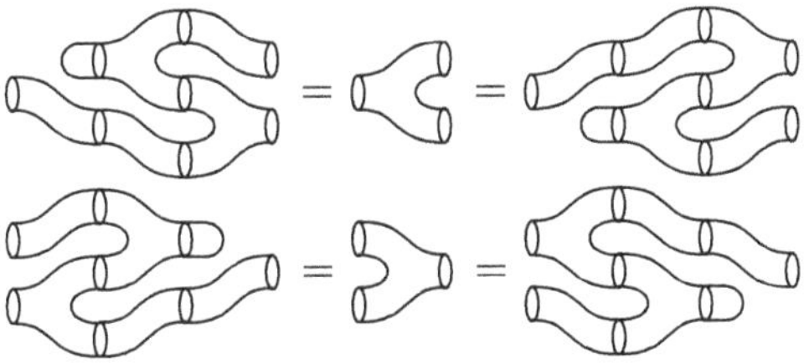

These are 2.3.16 and 2.3.17 above. The Frobenius relation is an immediate consequence, cf. 2.3.19. The nondegenerate pairing is mentioned explicitly, but the Frobenius relation is not.

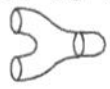

The first explicit mention of the Frobenius relation, and a proof of 2.3.24, were given in 1991 by Quinn [43], unaware of [32]. However, Quinn required the symmetry axiom

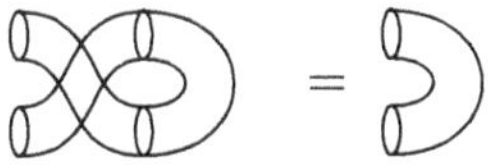

and called the algebras *ambialgebras* (although he was aware of the existing terminology *symmetric (Frobenius) algebra* (as employed in Curtis and Reiner [15])). Independently, Abrams gave the commutative case of the new characterisation in [1] (1995), and the noncommutative case appeared in [2] (1998).

Commutativity and cocommutativity

2.3.26 The twist map. For every pair of vector spaces V, V' there is a canonical twist map

$$\sigma_{V,V'} : V \otimes V' \to V' \otimes V$$

which simply changes the order of the factors. We picture the twist map like this:

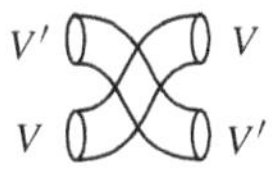

V' V

V V'

Note that now we have more than one vector space in play, so it is better to label the circles that represent these spaces.

The twist map satisfies some obvious axioms, among which the requirement $\sigma_{V,V'}\sigma_{V',V} = \text{id}_{V\otimes V'}$:

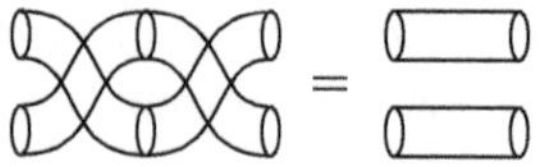

Taken together, the properties amount to saying that $\mathbf{Vect}_{\Bbbk}$ is a symmetric monoidal category, cf. 3.2.28.

Of course, we are particularly interested in the case where $V = V' = A$ is an algebra, so that we have the multiplication map  . Then naturality of the twist map with respect to this amounts to this relation:

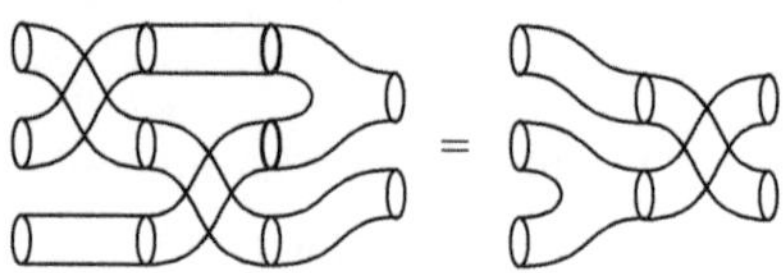

2.3.27 Commutative algebras. Let A be an algebra, with multiplication  . Then we can picture the axiom of being a commutative algebra:

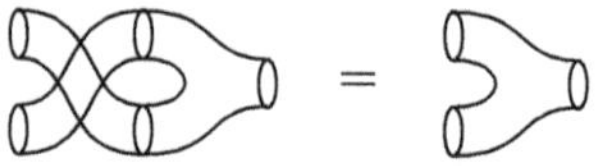

2.3.28 Cocommutative coalgebras. A coalgebra A with comultiplication is said to be *cocommutative* if this relation holds:

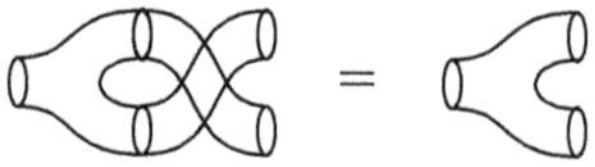

2.3.29 Proposition. *The comultiplication of a Frobenius algebra is cocommutative if and only if the multiplication is commutative.*

Proof. Suppose the multiplication is commutative. We will show that the map

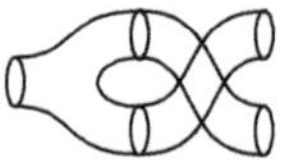

has ε as counit and satisfies the Frobenius relation. Then we can invoke the result (2.3.22) that such a comultiplication is unique in a Frobenius algebra, and thus conclude that $\delta\sigma = \delta$. The converse implication follows from duality.

So let us establish the left-hand side of the Frobenius relation, with the map $\delta\sigma$ as 'comultiplication', i.e. this equation:

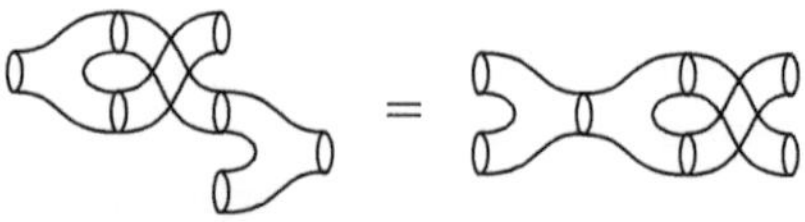

Here is the proof of that statement:

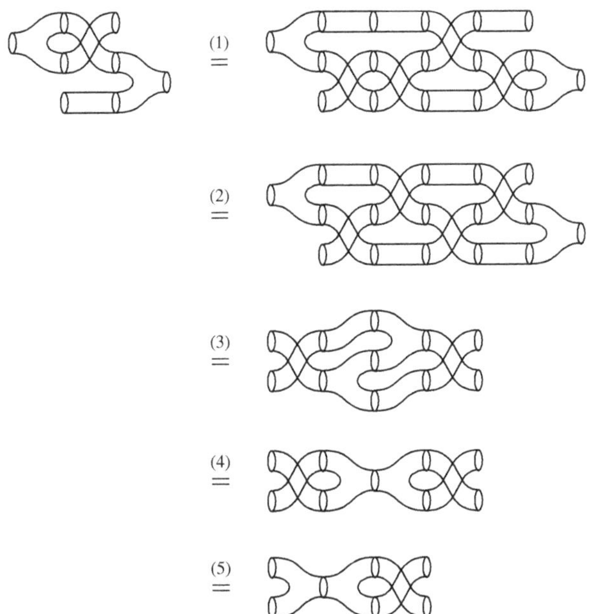

The first step was to insert three new twist maps: the two maps just give the identity, and the third map, inserted just before, is justified by commutativity. In step (2) we changed the order of the three rightmost twist maps, according to the 'symmetric-group-relation' – it is an instance of the naturality of the twist map. Equation (3) expressed another two instances of naturality, this time with respect to comultiplication and multiplication – the twist maps in the middle move outwards past comult and mult. Step (4) is the Frobenius relation, and finally in (5) we used commutativity of the multiplication map back again. □

2.3.30 Symmetric Frobenius algebras. If (A, β) is a Frobenius algebra then we can also picture the condition of being a symmetric one (that β satisfies the trace condition):

=

Tensor calculus (linear algebra in coordinates)

Until now we have carefully avoided coordinates. In this subsection we will write out everything in coordinates – not because we need to do so, but just for the fun of it. (This subsection will be used only in the exercises.) We adopt the elegant tensor element notation common in Riemannian geometry and physics, suppressing summation signs, and carefully distinguishing upper and lower indices. As it turns out, this tensor notation goes hand in hand with the graphical calculus.

Let A be as above. Fix a basis $\{T_0, \ldots, T_r\}$ for the vector space A, in such a way that $T_0 = 1_A$.

2.3.31 The multiplication tensor. Since the multiplication $\mu : A \otimes A \to A$ is linear we can describe it completely by specifying what it does to the elements of the basis. Recall that a canonical basis for $A \otimes A$ is $T_i \otimes T_j$ where i and j run from 0 to r. The result of multiplying two such basis element is then a linear combination of $T_0, \ldots, T_r$:

$$T_i \otimes T_j \mapsto T_k \mu^k_{ij}.$$

Here and throughout the section we suppress sum signs, adopting the Einstein summation convention: all repeated indices (which appear as one lower and one upper) are assumed to be summed. In this case the symbol k is repeated, so the meaning is $\sum_{k=0}^{r} T_k \mu^k_{ij}$.

We always write the indices corresponding to input as lower indices and the indices corresponding to output as upper indices. This is the usual convention from Riemannian geometry where we would call this a (2, 1)-tensor. We draw it like this

$$\mu^k_{ij} \; :$$

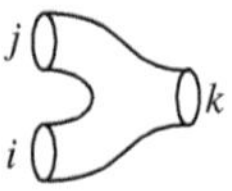

Note that the input holes correspond to lower indices and that the output corresponds to upper indices.

The associativity of μ can now be written explicitly. In the product $T_i T_j T_k$ we can either start by multiplying T_i and T_j, and then multiply the result with T_k:

$$T_i \otimes T_j \otimes T_k \mapsto T_e \mu^e_{ij} \otimes T_k \mapsto T_l \, \mu^e_{ij} \mu^l_{ek}$$

Or we can do it the other way around, which gives $T_l \, \mu^e_{jk} \mu^l_{ie}$. Now by linear algebra, two such linear combinations are equal if and only if all the coefficients are equal. So altogether the associativity equation is

$$\mu^e_{ij} \mu^l_{ek} = \mu^e_{jk} \mu^l_{ie}.$$

It is interesting to compare with the graphical representation:

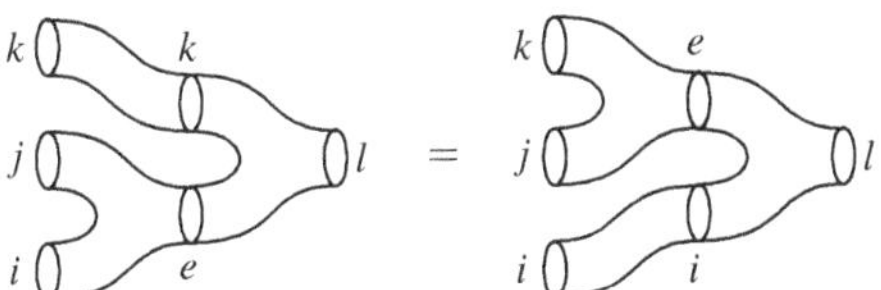

If you know the coordinate expression we just found, you can draw the picture just by putting together the pieces according to the matching indices. Or conversely, if we know the abstract associativity relation from 2.3.8 we can quickly find the tensor expression, by putting labels (indices) on all the holes, and then write down the tensor element of each piece.

2.3.32 The metric tensor. Since the bilinear pairing $\beta : A \otimes A \to \Bbbk$ is nondegenerate we will call it a *metric*, but of course, since we are over a general field $\Bbbk$, it makes no sense to talk about positive definiteness, as one usually requires in Riemannian geometry... We regard β as a $(2, 0)$-tensor. Its tensor elements are the constants $\beta_{ij} \in \Bbbk$ defined by

$$\beta_{ij} = \langle T_i \,|\, T_j \rangle .$$

We will draw this like this

β_{ij} :

2.3.33 Associativity of the metric tensor, and the three-point functions. The requirement that β be associative (cf. 2.1.32) is now easy to write down explicitly:

$$\mu_{ij}^{e} \beta_{ek} = \mu_{jk}^{f} \beta_{if} .$$

Now the coordinate expression for the three-point function (see 2.3.12) is

$$\phi_{ijk} := \mu_{ij}^{e} \beta_{ek} = \mu_{jk}^{f} \beta_{if}$$

See how the metric β_{ij} serves to lower indices. The important fact is that β is nondegenerate so we can take the inverse matrix and use it to raise indices.

2.3.34 Nondegeneracy. Let us write out in coordinates what it means to say that the pairing $\beta : A \otimes A \to \Bbbk$ is nondegenerate. To this end, let (ι_i^j) denote the identity matrix – in other words, ι_i^j is the 'Kronecker delta': equal to 1 for $i = j$ and equal to 0 otherwise. (The usual symbol δ for the Kronecker delta is currently assigned to the comultiplication. . .)

Now nondegeneracy of β (cf. 2.1.10 and the snake relation 2.3.11) has the following coordinate expression: *there exists a* $(0, 2)$*-tensor* $\gamma : \Bbbk \to A \otimes A$ *such that*

$$\beta_{ij}\gamma^{jk} = \iota_i^k \qquad \text{and} \qquad \gamma^{ij}\beta_{jk} = \iota_k^i .$$

In other words, it amounts to the statement that the matrix (β_{ij}) is invertible and that (γ^{ij}) is its inverse. We draw γ like this:

γ^{ij} :

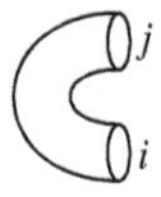

2.3.35 Multiplication in terms of the three-point function – coordinate expression of 2.3.13. Now that we have the matrix (γ^{ij}), we can use it to raise indices, and express the multiplication in terms of the three-point function.

$$\phi_{ije}\gamma^{ek} = \mu_{ij}^k = \gamma^{kf}\phi_{fij}$$

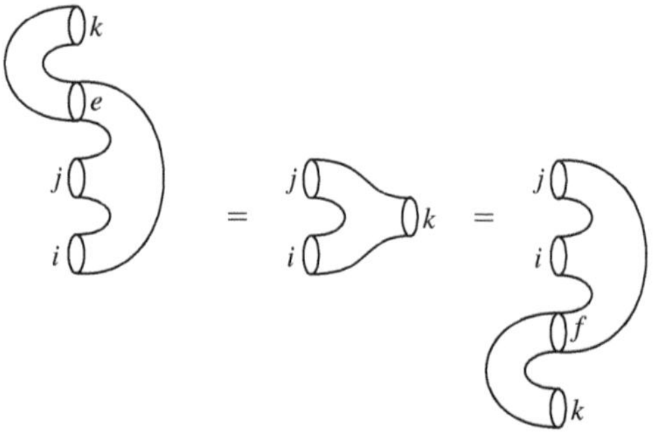

Proof (in coordinates): let us prove the left-hand equation. By definition, $\phi_{ije} = \mu_{ij}^s\beta_{se}$; now multiply this equation from the right with the matrix (γ^{ek}).

2.3.36 Comultiplication (cf. 2.3.16). Define the comultiplication to be

$$\begin{aligned} A &\longrightarrow A \otimes A \\ T_k &\longmapsto T_i \otimes T_j\delta_k^{ij}, \end{aligned}$$

where the tensor elements are given by

$$\delta_k^{ij} := \gamma^{ie}\mu_{ek}^j = \mu_{kf}^i\gamma^{fj}.$$

For this to make sense we must prove that these two expressions agree. This follows from 2.3.35: both sides are equal to

$$\gamma^{ie}\phi_{ekf}\gamma^{fj}.$$

Conversely, we can express the multiplication tensor in terms of the comultiplication, by lowering indices:

$$\mu^k_{ij} = \beta_{ie}\delta^{ek}_j = \delta^{kf}_i \beta_{fj}.$$

2.3.37 The Frobenius condition. The coordinate expression of 2.3.19 is

$$\delta^{em}_j \mu^l_{ie} = \mu^e_{ij}\delta^{lm}_e = \delta^{le}_i \mu^m_{ej}.$$

Let us prove the left-hand equation; the corresponding picture is

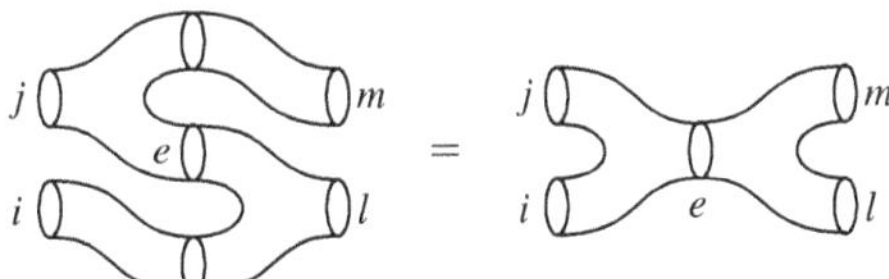

The coordinate proof is the same as the graphical proof we gave for 2.3.19:

$$\delta^{em}_j \mu^l_{ie} = \gamma^{km}\mu^e_{jk}\mu^l_{ie} = \gamma^{km}\mu^e_{ij}\mu^l_{ek} = \mu^e_{ij}\delta^{lm}_e.$$

Here the first step was to use the definition of the comultiplication (cf. 2.3.36); then use the associativity equation 2.3.31, then 2.3.36 backwards.

2.3.38 Example. Now that we are comfortable with coordinates, let us write out the comultiplication in the Frobenius algebra $\mathbb{C}$ of Example 2.2.14. So take this basis:

$$T_0 := 1 \qquad T_1 := \sqrt{-1}.$$

Then the tensor elements for the multiplication are

$$\mu^0_{00} = 1, \quad \mu^1_{01} = \mu^1_{10} = 1, \quad \mu^0_{11} = -1$$

(those not listed are equal to zero). Our Frobenius form is by definition

$$T_0 \mapsto 1$$
$$T_1 \mapsto 0,$$

so the Frobenius pairing is given by

$$\beta_{00} = 1, \quad \beta_{01} = \beta_{10} = 0, \quad \beta_{11} = -1.$$

The inverse of this matrix $\beta = \left(\begin{smallmatrix} 1 & 0 \\ 0 & -1 \end{smallmatrix}\right)$ is $\gamma := \left(\begin{smallmatrix} 1 & 0 \\ 0 & -1 \end{smallmatrix}\right)$. Using the definition of the comultiplication, $\delta_k^{ij} = \gamma^{ie}\mu_{ek}^{j}$, we find

$$\delta_0^{00} = 1, \quad \delta_0^{11} = -1, \qquad \delta_1^{01} = \delta_1^{10} = 1;$$

(the others are zero). So the comultiplication is given by

$$\begin{aligned} T_0 &\mapsto T_0 \otimes T_0 - T_1 \otimes T_1 \\ T_1 &\mapsto T_0 \otimes T_1 + T_1 \otimes T_0, \end{aligned}$$

just like in Example 2.3.3. In conclusion, *Sweedler's trigonometric coalgebra 2.3.3 over the real numbers is nothing but the Frobenius algebra structure on* $\mathbb{C}$.

Exercises

1. Complete the proof of 2.3.17.
2. Prove Lemma 2.3.21.
3. Decorate the snake relation with indices in accordance with the coordinate expression of 2.3.34.

The remaining exercises centre around the handle operator  . If you look back at the classification of surfaces done in 1.4.15–1.4.16 you see that this is really the most interesting piece topologically – it is 'where the genus is'! Algebraically it also has some important properties – and anyway is a good excuse to do some computations...

4. *The handle operator* is the $\Bbbk$-linear map $\omega : A \to A$ defined as the composite $\delta\mu$ – so what you actually want to know is this picture:  . Show that if A is a Frobenius algebra then $\omega : A \to A$ is a right (and left) A-module homomorphism, i.e.

$$\begin{array}{ccc} A \otimes A & \xrightarrow{\omega \otimes \mathrm{id}_A} & A \otimes A \\ {\scriptstyle \mu}\downarrow & & \downarrow{\scriptstyle \mu} \\ A & \xrightarrow[\omega]{} & A \end{array}$$

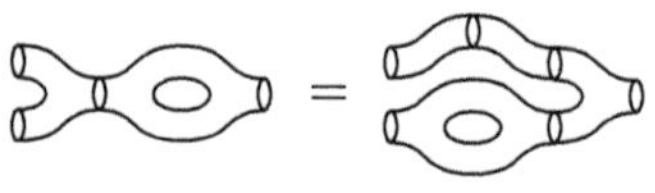

=

of course you are expected to do the proof graphically!

5. Show that the handle operator is given by multiplication by a central element w. This element is called the *handle element*. In the coordinate notation of 2.3.34, $w = T_i \gamma^{ij} T_j$. (While it is very easy to see that w is central from a graphical argument, it can be messy to prove this with coordinates!)

⋆6. Show that if the handle element belongs to Null(ε) then it is zero.
7. Show that if (A, ε) is a symmetric Frobenius algebra over a field of characteristic zero, then the handle element w is nonzero. (Hint: it may actually be easier to prove the stronger result that $w\varepsilon = \dim A$, using Exercise 10 on page 34.)
8. Use coordinates to compute the handle operator in the Frobenius $\mathbb{R}$-algebra $\mathbb{C} \simeq \mathbb{R}[t]/(t^2+1)$ with Frobenius form 'taking the real part'. Most of this computation was already performed in Example 2.3.38, and the answer is already known from the previous exercise, so this one mainly serves as model for the next couple of exercises.
9. Consider the group algebra of $G = \mathbb{Z}/2\mathbb{Z}$. It is $\Bbbk[t]/(t^2-1)$, and as Frobenius form we take $1 \mapsto 1$, $t \mapsto 0$. Compute the coordinate expressions of pairing, multiplication, copairing, comultiplication, and handle operator, with respect to this basis. (Observe that if the characteristic of $\Bbbk$ is 2, then the handle operator is zero. Otherwise it is invertible.)
10. Consider yet a third Frobenius algebra of dimension 2: the algebra $A = \Bbbk[t]/t^2$ (in algebraic geometry it is called the ring of dual numbers; it is also the cohomology ring of $\mathbb{CP}^1$). This is a Frobenius algebra by 2.2.21 (and 2.2.23), $\varepsilon : A \to \Bbbk$, $t \mapsto 1$, $1 \mapsto 0$. Take a basis $T_0 = 1$, $T_1 = t$, and work out the coordinate expression of multiplication, pairing, copairing, comultiplication, and handle operator, just as it was done in 2.3.38 to show that the comultiplication is given by

$$\begin{aligned} T_0 &\mapsto T_0 \otimes T_1 + T_1 \otimes T_0 \\ T_1 &\mapsto T_1 \otimes T_1. \end{aligned}$$

Show that the handle operator $\delta\mu : A \to A$ has square zero.
11. It was a coincidence that the exponent 2 in the previous exercise matched the 'square-zero' conclusion. To be sure, work out the exercise again with $A = \Bbbk[t]/t^n$, for general $n > 1$. The conclusion is still that the handle operator has square zero.
12. Noncommutative example. Consider the algebra $\Bbbk\langle x, y\rangle/(x^2, y^2, xy + yx)$, with basis $\{1, x, y, xy\}$. Show this is a Frobenius form:

$$\begin{aligned} xy &\mapsto 1 \\ \text{others} &\mapsto 0. \end{aligned}$$

(The best way is probably to write down the whole multiplication table and check that the induced pairing is nondegenerate – you will need this information anyway.) Use coordinates as in the exercises above to show that the handle element is zero.

13. Yet another little algebra to play with: $\Bbbk\langle x, y\rangle/(x^2+1, y^2-1, xy+yx)$. At first, this looks a lot like the previous example... Show that there are no nilpotent elements in this ring (in particular it is semi-simple). Find a Frobenius form with respect to which the handle element is nonzero, and find another Frobenius form with respect to which the handle element is zero.
14. Work out the group algebra of the symmetric group on three letters, $\Bbbk\langle x, y\rangle(x^2-1, y^2-1, xyx-yxy)$, in the basis $\{1, x, y, xy, yx, xyx\}$. First take as Frobenius form the one described in 2.2.18, ($1 \mapsto 1$, others mapping to zero), and show that the handle element is 6. Next take as Frobenius form $xyx \mapsto 1$, others mapping to zero. Show that the handle element is $2(xyx + x + y)$.

⋆15. Let A be a commutative local Frobenius algebra, i.e. an artinian Gorenstein ring, cf. 2.2.21. Show that the handle element generates the socle (as ideal, and also as $\Bbbk$-vector space).

16. Consider a direct product algebra $A = A' \times A''$ with Frobenius form $\varepsilon = \varepsilon' + \varepsilon''$ (where ε' and ε'' are Frobenius forms on A' and A'', as in Exercise 7 on page 105). Show that the handle operator of A is the direct product of the handle operators of A' and A''.

⋆17. Let (A, ε) be a Frobenius algebra. Show that the handle element annihilates the Jacobson radical. (This is a bit technical, but the next two exercises give easy corollaries to this result.)
Here is the strategy (cf. Sawin [44] for the commutative case). Let $\mathfrak{a}_1$ denote the right annihilator of the Jacobson radical $J(A)$. Define inductively a chain of 2-sided ideals $\mathfrak{a}_1 \subset \mathfrak{a}_2 \subset \cdots \subset \mathfrak{a}_r = A$, letting $\mathfrak{a}_{k+1}$ be the right annihilator of the right A-module $J(A/\mathfrak{a}_k)$ (Jacobson radical of the quotient $A/\mathfrak{a}_k$). Pick a basis for $\mathfrak{a}_1$, expand to a basis for $\mathfrak{a}_2$ and so on, getting a basis $\{t_1, \ldots, t_n\}$ for A. Consider the right-dual basis $\{\check{t}_1, \ldots, \check{t}_n\}$, where $\check{t}_i := \gamma^{ie} t_e$; this means that $\langle\, t_i \,|\, \check{t}_j \,\rangle = \iota_i^j$, in the notation of 2.3.34. Now suppose $t_i \in \mathfrak{a}_{k+1} \smallsetminus \mathfrak{a}_k$, and let $x \in J(A)$. Then $xt_i \in \mathfrak{a}_k$, so by our choice of basis and dual basis we have $0 = \langle\, xt_i \,|\, \check{t}_i \,\rangle = (xt_i\check{t}_i)\varepsilon$. Hence by the Frobenius condition the left ideal $J(A)t_i\check{t}_i$ is zero, and hence $t_i\check{t}_i \in \mathfrak{a}_1$. This argument holds for each i, so in conclusion the handle element $w = \sum_i t_i\check{t}_i$ annihilates $J(A)$.

18. Let (A, ε) be a Frobenius algebra. Show that if the handle element w is nilpotent then $w^2 = 0$. (Hint: $J(A)$ contains every nilpotent left ideal.)
19. (Cf. Abrams [3] for the commutative case.) Show that if the handle element is invertible then A is semi-simple.

20. Let A be a commutative and semi-simple Frobenius algebra. (Then it is a direct product of fields, by Wedderburn's theorem (see 2.2.17), and since every simple commutative ring is a field.) Show that the handle element is invertible.
21. Construct an example of a four-dimensional Frobenius algebra whose handle element is neither invertible nor of square zero. (Hint: use some of the 2-dimensional examples above together with Exercise 16.)
⋆22. (Cf. Abrams [3].) If you are acquainted with cohomology rings (cf. 2.2.23). Let X be a compact connected orientable manifold of dimension r, and put $A = H^*(X)$. Show that the Euler class (top Chern class of the tangent bundle, $c_r(T_X)$) is the handle element of A.

2.4 The category of Frobenius algebras

Frobenius algebra homomorphisms

2.4.1 Duality. It is particularly clear from the pictures that there is a complete symmetry between μ and η on one side and δ and ε on the other side. As a consequence, if $(A, \eta, \mu, \delta, \varepsilon)$ is a Frobenius algebra then the dual vector space A^*, becomes a Frobenius algebra again, by taking the 'duals' of $\eta, \mu, \delta, \varepsilon$ as structure maps. There is a subtlety, however: the dual of $\delta : A \to A \otimes A$ is actually a map $(A \otimes A)^* \to A^*$, so in order to get a true multiplication map on A^* we need to compose with the canonical isomorphism $\psi : A^* \otimes A^* \xrightarrow{\sim} (A \otimes A)^*$ described in 2.1.17. For this reason it is not a completely trivial fact that this new multiplication map is associative. It must be checked that the isomorphisms ψ are compatible with iterated tensor products. (When you check this, remember that $(\delta \otimes \mathrm{id})^* = \mathrm{id} \otimes \delta^*$, cf. 2.1.17...)

Similarly the dual of $\mu : A \otimes A \to A$ is actually a map $A^* \to (A \otimes A)^*$ which we must compose with the inverse of ψ in order to get a true comultiplication map.

2.4.2 Lemma. *If a $\Bbbk$-algebra homomorphism ϕ between two Frobenius algebras (A, ε) and (A', ε') is compatible with the forms in the sense that the diagram*

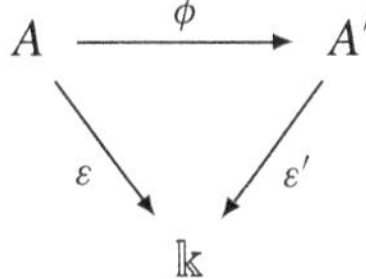

commutes, then ϕ is injective.

(Being a $\Bbbk$-algebra homomorphism means that it is multiplicative: $ab\phi = a\phi\, b\phi$ and respects the units: $1\,\phi = 1'$.)

Proof. The kernel of ϕ is an ideal and it is clearly contained in $\mathrm{Null}(\varepsilon)$. But $\mathrm{Null}(\varepsilon)$ contains no nontrivial ideals, so $\mathrm{Null}(\phi) = 0$ and thus ϕ is injective. □

2.4.3 Example. Let $\mathbb{R}$ be the trivial Frobenius algebra over $\mathbb{R}$, and let $\mathbb{C}$ be the Frobenius algebra of Example 2.2.13: the Frobenius form $\mathbb{C} \to \mathbb{R}$ is 'taking the real part'. The canonical injection $\mathbb{R} \hookrightarrow \mathbb{C}$ is compatible with the Frobenius forms, but not with the comultiplication (cf. the coordinate description given in 2.3.38).

2.4.4 The category of Frobenius algebras. A *Frobenius algebra homomorphism* $\phi : (A, \varepsilon) \to (A', \varepsilon')$ between two Frobenius algebras is an algebra homomorphism which is at the same time a coalgebra homomorphism. In particular it preserves the Frobenius form, in the sense that $\varepsilon = \phi\varepsilon'$. Let $\boldsymbol{FA}_{\Bbbk}$ denote the category of Frobenius algebras over $\Bbbk$ and Frobenius algebra homomorphisms, and let $\boldsymbol{cFA}_{\Bbbk}$ denote the full subcategory of all commutative Frobenius algebras.

2.4.5 Lemma. *A Frobenius algebra homomorphism $\phi : A \to A'$ is always invertible. (In other words, the category $\mathbf{FA}_{\Bbbk}$ is a groupoid (and so is $\mathbf{cFA}_{\Bbbk}$).)*

Proof. Since ϕ is comultiplicative and respects the counits ε and ε' (as well as the units η and η'), the 'dual' map $\phi^* : A'^* \to A^*$ is multiplicative and respects units and counits. But then the preceding lemma applies and shows that ϕ^* is injective. Since A is a finite-dimensional vector space this implies that ϕ is surjective. We already know it is injective, hence it is invertible. □

Tensor products of Frobenius algebras

2.4.6 Tensor products of algebras. Given two algebras A and A', consider their tensor product $A \otimes A'$ as vector spaces. Now component-wise multiplication makes $A \otimes A'$ into an algebra:

$$\begin{aligned} (A \otimes A') \otimes (A \otimes A') &\longrightarrow A \otimes A' \\ (x \otimes x') \otimes (y \otimes y') &\longmapsto xy \otimes x'y'. \end{aligned}$$

Note that A only interacts with A, and A' only with A', and that the twist map is crucial in order to construct the new multiplication map from the existing maps. This is particularly clear from the graphical version of the multiplication map on $A \otimes A'$.

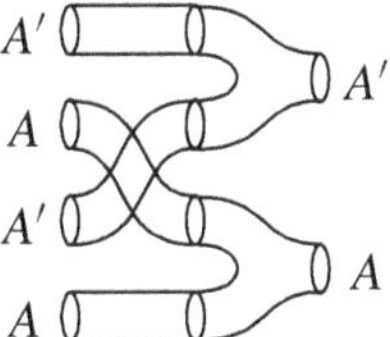

Now let us check that this new multiplication map is associative (knowing that the two multiplication maps on A and A' are so). It is quite easy to do that just by writing down the equations in terms of elements, but for the fun of it we will do the graphical proof (which is also easy, although it displays the rôle of the twist map which is hidden in the element-equation version). The graphical version of this claim is

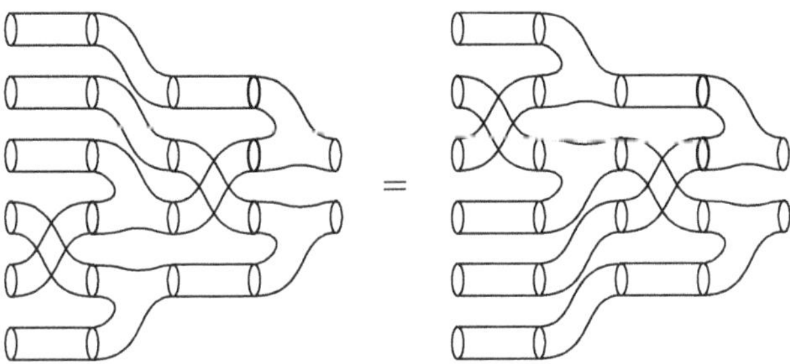

And here is the proof: on the left-hand side of the equation, begin by moving the second twist map over to the left of the adjacent multiplication map, and also move the lower multiplication a bit to the right

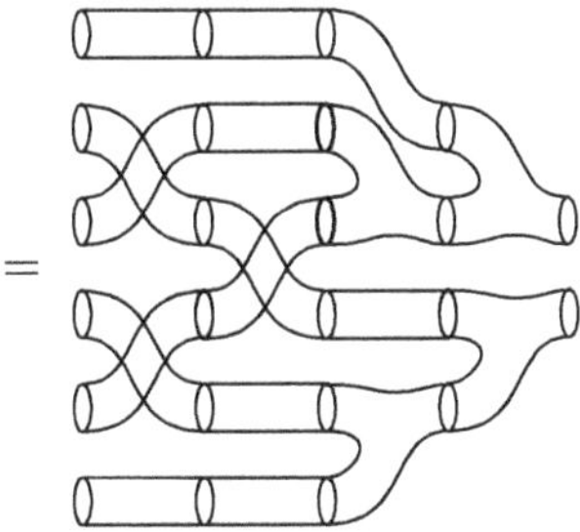

Now apply the associativity equation 2.3.8 twice (once for A and once for A') to arrive at

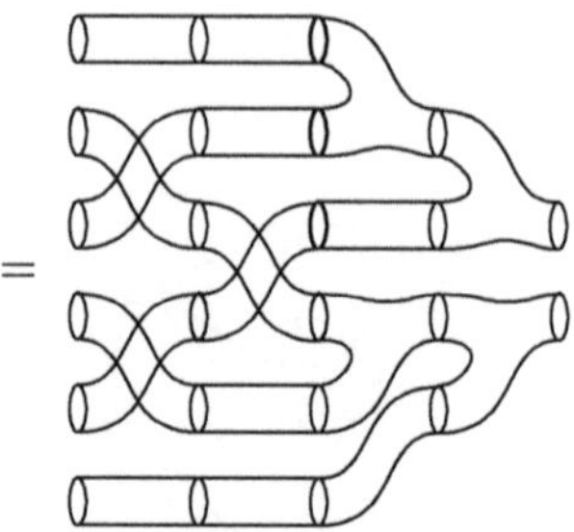

Finally use the naturality of the twist to move the two lower twist maps to the right of the multiplication map they precede. This move leads to the right-hand side of the asserted equation.

In order for $A \otimes A'$ to be an algebra, of course we should also specify the unit map. It is simply

$$\begin{aligned} \Bbbk &\longrightarrow A \otimes A' \\ 1 &\longmapsto 1_A \otimes 1_{A'} \end{aligned}$$

which is pictured

A'

A

It is easy to check the unit axioms.

2.4.7 Tensor products of coalgebras. Now it is an easy exercise to show that the tensor product of two coalgebras is again a coalgebra in a natural way. The figures are just the mirror images of those above.

Finally,

2.4.8 The tensor product of two Frobenius algebras is in a natural way again a Frobenius algebra. Here of course we use the characterisation of Frobenius algebras in terms of comultiplication and the Frobenius condition (2.3.24). We already know that the tensor product is again an algebra and a coalgebra. It remains to show that the Frobenius condition holds:

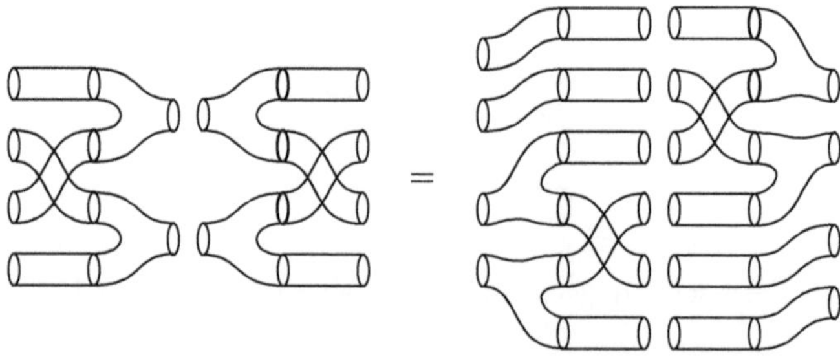

(This is only the right-hand part of the equation. . .)

Here is the proof: on the left-hand side of the equation, begin by using the Frobenius relation (right-hand part) on the two occurrences of (one for A and one for A'), to get

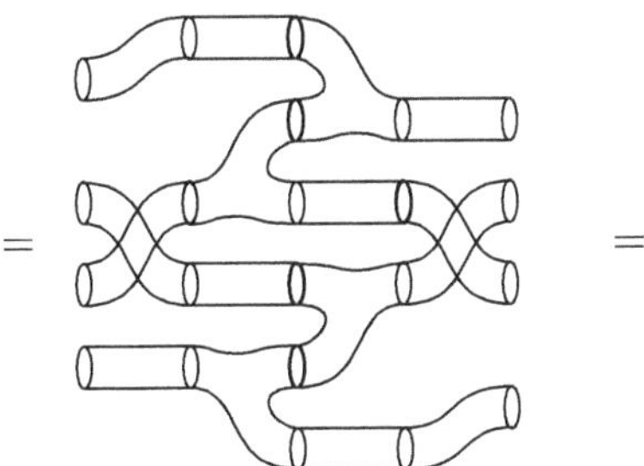

Now use the naturality of the twist map to move the two twist maps towards the centre of the picture (they become four instead of two, but two of them cancel out each other):

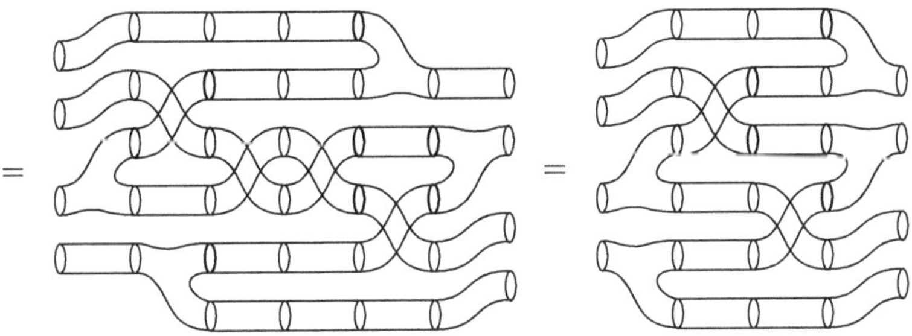

which is what we claimed (modulo a last move where the upper twist moves right and the lower one moves left).

(In fact, since the Frobenius condition (together with unit and counit conditions) implies associativity and coassociativity, there was no need to check associativity and coassociativity separately. . .)

So $\boldsymbol{FA}_{\Bbbk}$ is closed under tensor products, and we also noted in 2.2.12 that $(\Bbbk, \mathrm{id}_{\Bbbk})$ is a Frobenius algebra. In fact, $(\boldsymbol{FA}_{\Bbbk}, \otimes, \Bbbk)$ is a monoidal category, cf. 3.2.

Digression on bialgebras

(This subsection is not really needed elsewhere in these notes.)

2.4.9 Bialgebras. There is another very common sort of algebra which is simultaneously an algebra and coalgebra, usually called a bialgebra (see Kassel [29], Chapter III). Observe that bialgebras can be of infinite dimension.

Bialgebras are also characterised by a compatibility condition, but this time instead of requiring that δ be A-linear, we require it to be an A-algebra homomorphism. This makes sense since we have now explained what the canonical algebra structure on $A \otimes A$ is. In fact we also require ε to be an algebra homomorphism.

Precisely, there are four conditions. Two conditions amount to the statement that δ is a $\Bbbk$-algebra homomorphism:

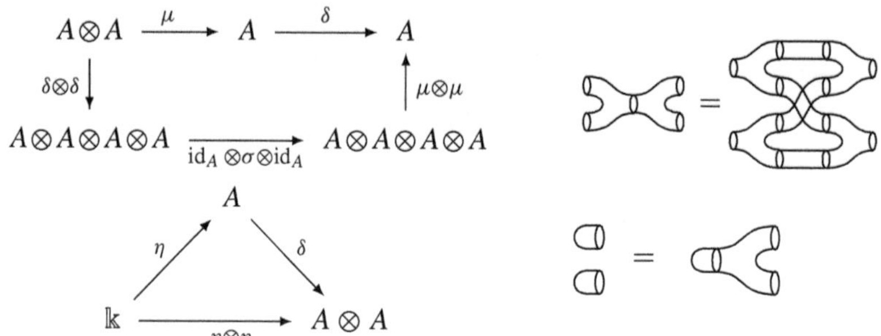

And two conditions amount to the statement that ε is a $\Bbbk$-algebra homomorphism:

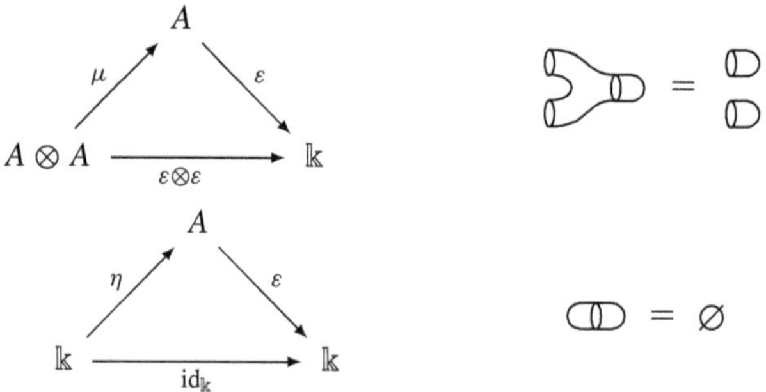

We see that these compatibility conditions are very different from those that define a Frobenius algebra. In fact,

2.4.10 Proposition. *Let A be a bialgebra over $\Bbbk$ of finite dimension, with structure maps $\eta, \mu, \delta, \varepsilon$ as above. If ε is a Frobenius form then $A \simeq \Bbbk$.*

Proof. The conditions $\mu\varepsilon = \varepsilon \otimes \varepsilon$ and $\eta\varepsilon = \mathrm{id}_{\Bbbk}$ express that ε is a ring homomorphism, so in particular $\mathrm{Null}(\varepsilon)$ is an ideal in A. If ε is furthermore a Frobenius form then this ideal must be the zero ideal, so $A \simeq \Bbbk$. □

2.4.11 Hopf algebras. A particularly important class of bialgebras is the Hopf algebras. A *Hopf algebra* is a bialgebra equipped with a $\Bbbk$-linear map $S : A \to A$ called an antipode (pictured ⊳⊲) with the following properties:

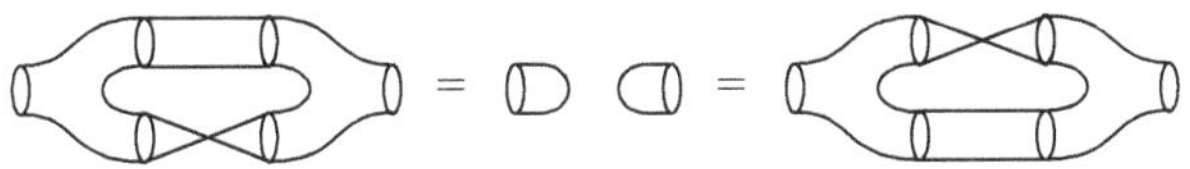

The symbol $\bowtie$ is meant to indicate reversal or reflection. It is not meant to indicate that there is a topological surface of any kind that can realise it – in fact we have already seen that the axioms for a bialgebra are not at all topological.

2.4.12 Group algebras as Hopf algebras or Frobenius algebras. Group algebras are naturally Hopf algebras, and thus bialgebras. As just stated in the Proposition, the counit of a bialgebra can never be a Frobenius form, so here we are apparently at odds with Example 2.2.18 where we showed that group algebras are Frobenius algebras. This is not a contradiction – the point is that we are talking about different linear functionals. The counit of a group algebra considered as a bialgebra is the one that sends each generator to 1. The Frobenius form of Example 2.2.18 was $1 \mapsto 1$, other generators map to 0. In particular, the coalgebra structure determined by a Frobenius form on a group algebra is *not* the coalgebra structure coming from the diagonal map (cf. 2.3.2).

On the other hand, one can show that every finite-dimensional Hopf algebra admits a Frobenius algebra structure – see Sweedler [46], Chapter V. (So this generalises 2.2.18.)

Exercises

1. 1-dimensional Frobenius algebras. We saw in 2.2.12 that the $\Bbbk$-algebra $\Bbbk$ with linear functional $\mathrm{id} : \Bbbk \to \Bbbk$ is a Frobenius algebra. We also noted that any nonzero linear map $\Bbbk \to \Bbbk$ will do as Frobenius form. Show that all these Frobenius structures are nonisomorphic.

That was the only exercise in this section. Make sure you did all the exercises in the previous section, where there were quite a few...

3

Monoids and monoidal categories

Summary

After having flirted with categorical viewpoints throughout the first two chapters, we now go the whole length. (The reader can consult the short appendix for the notions of categories, functors, and natural transformations, if necessary.)

The chapter starts with a warm-up section on monoids. Here we take an 'advanced viewpoint' on some very simple things, and prepare ourselves for some less trivial mathematics.

In Section 3.2, monoidal categories are introduced – always considered to be strict. (A short interlude takes a look at the nonstrict case and states Mac Lane's coherence theorem.) A monoidal category is a category with a 'multiplication' and a 'neutral object', satisfying certain natural axioms. Crucial examples are: $(\textbf{\textit{Vect}}_{\Bbbk}, \otimes, \Bbbk)$, $(\textbf{\textit{Set}}, \coprod, \emptyset)$, and also $(\textbf{\textit{2Cob}}, \coprod, \emptyset)$. The notions of symmetric monoidal category and symmetric monoidal functor are introduced.

In Chapter 1 we defined a 2-dimensional TQFT to be a symmetric monoidal functor from ***2Cob*** to $\textbf{\textit{Vect}}_{\Bbbk}$. Such functors form a category, the arrows being monoidal natural transformations. Now we prove the Main Theorem (3.3.2): *the category of 2D TQFTs is equivalent to the category of commutative Frobenius algebras.*

The second half of this chapter aims at placing this result in its proper context. To prepare for this, Section 3.4 is devoted to the study of two important monoidal categories: the category of finite ordinals $\Delta = \{\mathbf{0}, \mathbf{1}, \mathbf{2}, \dots\}$ (simplex category), and the category Φ of finite cardinals (symmetric version of Δ). Several different descriptions of these categories are given, notably a graphical interpretation, and presentation in terms of generators and relations.

Section 3.5 studies monoids in monoidal categories. (For instance, a monoid in $\textbf{\textit{Vect}}_{\Bbbk}$ is precisely a $\Bbbk$-algebra.) The object $\mathbf{1}$ is a monoid in Δ, and it is shown to have the following universal property: *every monoid in any monoidal category* **V** *is the image of* **1** *under a unique monoidal functor* $\Delta \to \mathbf{V}$.

Finally in Section 3.6 we copy over these constructions and arguments, introducing the concept of Frobenius object in a monoidal category (such that Frobenius objects in $\mathbf{Vect}_{\Bbbk}$ are precisely Frobenius algebras). The proof of the Main Theorem now carries over to establish the more general result (generalised main theorem (3.6.19)): *every commutative Frobenius object in a symmetric monoidal category* **V** *is the image of* **1** *under a unique symmetric monoidal functor* **2Cob** $\to$ **V**.

Monoidal categories were introduced in the 1960s by Bénabou, Mac Lane, and others (see the references and historical notes in Mac Lane's book [34]). The equivalence of categories between 2D TQFTs and commutative Frobenius algebras was discovered by Dijkgraaf [16] (1989), and more detailed proofs were provided by Quinn [43] and Abrams [1]. (Their treatments are sloppy with respect to the question of symmetry, however.) The characterisation of Δ as free monoidal category on a monoid seems to be due to Lawvere [32] (1967). The fact that **2Cob** is the free symmetric monoidal category on a commutative Frobenius object has been known for some years, but I do not know of a reference for this.

3.1 Monoids (in *Set*)

Some notions from set theory

3.1.1 Cartesian products. Given two sets X and Y, the *cartesian product* $X \times Y$ is the set of (ordered) pairs (x, y), where $x \in X$ and $y \in Y$. Similarly, for three sets X, Y, Z we have the notion of the cartesian product $X \times Y \times Z$, which is the set of (ordered) triples (x, y, z) where $x \in X$, $y \in Y$, and $z \in Z$. We identify $X \times Y \times Z$ with $(X \times Y) \times Z$ and also with $X \times (Y \times Z)$.

In general if we have n sets $X_1, \ldots, X_n$ then we have the n-fold cartesian product $X_1 \times \cdots \times X_n$ consisting of n-tuples $(x_1, \ldots, x_n)$ such that $x_i \in X_i$ for $i = 1, \ldots, n$. Again we identify it with any cartesian product obtained from setting parentheses.

An important case is when all the factors are identical: we write

$$X^n := \underbrace{X \times \cdots \times X}_{n \text{ factors}}.$$

The parentheses-deletion convention then amounts to

$$X^m \times X^n = X^{m+n}.$$

The 1-fold product $(X) = X^1$ we identify with X itself. (We may note also that $\times$ is in fact a functor, so given two maps $f : X \to Y$ and $f' : X' \to Y'$ there is a product map $f \times g : X \times Y \to X' \times Y' \dots$)

3.1.2 The singleton set. Important is the product $() = X^0$ without any factors! (Here the set X is arbitrary since anyway there are none of them in the product...) By the above axioms, we have $() \times (X_1 \times \cdots \times X_n) = X_1 \times \cdots \times X_n$, and in particular

$$() \times X = X = X \times ().$$

Since () is necessarily a singleton set, we will always denote it 1.

The singleton set 1 plays an important rôle as a device for singling out elements of a set. Precisely, given a set S there is an obvious canonical bijection

$$S \leftrightarrow \textbf{\textit{Set}}(1, S).$$

(Here $\textbf{\textit{Set}}(1, S)$ denotes the set of set maps from 1 to S.) Thus, instead of looking at elements we can look at arrows. Several variations will be made on this innocent looking theme (cf. 3.1.15, 3.5.18), and we will see that the principle is important.

3.1.3 Symmetry or twist. For each pair of sets (X, Y) there is a *twist map*

$$\begin{aligned} \mathrm{twist}_{X,Y} : X \times Y &\longrightarrow Y \times X \\ (x, y) &\longmapsto (y, x). \end{aligned}$$

Clearly, the composite $\mathrm{twist}_{X,Y}\,\mathrm{twist}_{Y,X}$ is equal to the identity. In general, given n sets $X_1, \dots, X_n$ there is a permutation map for each element of the symmetric group $\mathfrak{S}_n$. These maps compose just like the permutations compose in $\mathfrak{S}_n$. Note that all these permutations can be obtained by composing twist maps, according to 1.4.2.

The most important twist map for us will be the case

$$\begin{aligned} \mathrm{twist}_{X,X} : X \times X &\longrightarrow X \times X \\ (x, x') &\longmapsto (x', x). \end{aligned}$$

Clearly this is *not* the identity map of $X \times X$ – unless $X = 1$ or $X = \varnothing$.

Definition of monoid

3.1.4 Monoids. A monoid is a set M with a binary operation (composition law) which is associative and has a neutral element. (Exercise 3 shows the neutral element is unique if it exists.) If we employ infix notation for the composition (with a dot as infix), writing $(a, b) \mapsto a.b$, then the associativity axiom states that for every three elements a, b, c in M we have $(a.b).c = a.(b.c)$. The neutral element is an element $e \in M$ such that for all $a \in M$ we have $e.a = a = a.e$.

It is useful to express this in terms of commutative diagrams: A *monoid* is a set M together with two functions

$$\mu : M \times M \to M, \qquad \eta : 1 \to M$$

such that these three diagrams commute:

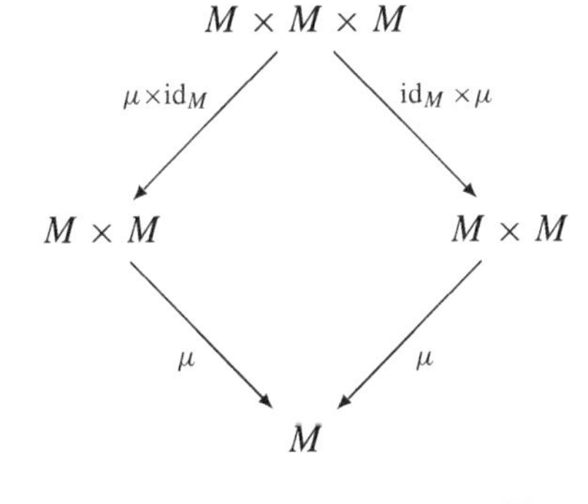

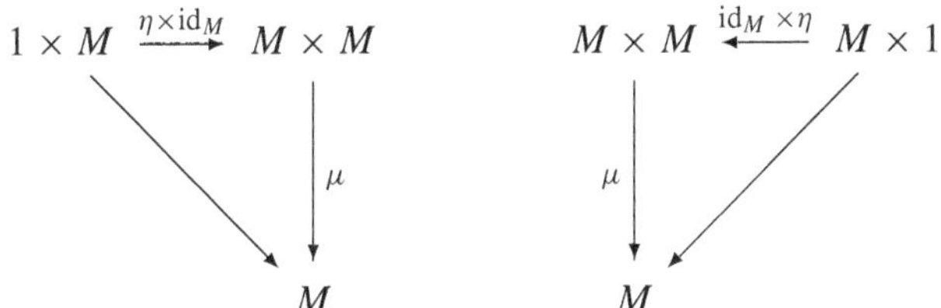

The symbols id_M stand for the identity function $M \to M$, and 1 stands for the singleton set. The diagonal maps without label are the canonical identifications.

We will refer to such a monoid by writing the triple (M, μ, η) or $(M, \,.\,, e)$.

3.1.5 ***n*-ary products.** Associativity implies that we can write and erase parentheses as we please, also in products involving more than three factors. In particular there are induced multiplication maps with more factors

$$\mu^{(n)} : M^n \to M, \qquad n \geq 2.$$

Just for completeness, let us furthermore define $\mu^{(1)} : M^1 \to M$ to be the identity map $\mathrm{id}_M : M \to M$, and define $\mu^{(0)} : M^0 \to M$ to be the neutral element map $\eta : 1 \to M$.

3.1.6 Monoid homomorphisms. A *monoid homomorphism* $\phi : M \to M'$ is a function that commutes with all the structure. Precisely,

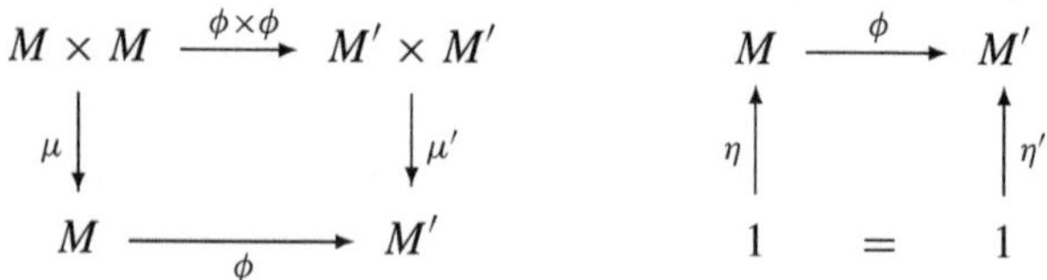

So in terms of compositions we have

$$(a\phi).(b\phi) = (a.b)\phi \quad \text{and also} \quad e\phi = e'.$$

We could also write this in a more uniform way, stating that a monoid homomorphism is a function that commutes with all $\mu^{(n)}$:

$$\begin{array}{ccc} M^n & \xrightarrow{\phi^n} & M^n \\ {\scriptstyle \mu^{(n)}}\downarrow & & \downarrow{\scriptstyle \mu^{(n)}} \\ M & \xrightarrow[\phi]{} & M \end{array} \qquad \text{for all } n \geq 0.$$

3.1.7 The category of monoids. One easily checks that the composition of two monoid homomorphisms is again a monoid homomorphism, and that the identity map is a monoid homomorphism, so altogether: there is a category denoted ***Mon*** whose objects are the monoids and whose arrows are the monoid homomorphisms. We write ***Mon***(X, Y) for the set of monoid homomorphisms from X to Y. A monoid homomorphism is called an *isomorphism of monoids* if there exists a two-sided inverse in ***Mon***. (Exercise 2 shows that the isomorphisms are precisely the bijective monoid homomorphisms.)

3.1.8 The product of two monoids. If M and M' are two monoids, then the product set $M \times M'$ has a canonical monoid structure, namely the one given by component-wise multiplication. That is, the multiplication on $M \times M'$ is given by

$$\begin{aligned} (M \times M') \times (M \times M') &\longrightarrow M \times M' \\ \big((x, x'), (y, y')\big) &\longmapsto (x.y, x'.y'). \end{aligned}$$

The unit map $1 \to M \times M'$ is simply the product of the two unit maps $1 \to M$ and $1 \to M'$.

Similarly, one could define for n monoids $M_1, \ldots, M_n$ a canonical monoid structure on the n-fold product $M_1 \times \cdots \times M_n$. (The empty product $(n = 0)$

would be the trivial monoid structure on 1 where both structure maps are the identity map.)

3.1.9 Commutative monoids. A monoid $(M, \ .\ , 1)$ is called *commutative* if for all elements a, b we have $a.b = b.a$. In terms of arrows and diagrams: a monoid is commutative if the multiplication $\mu : M \times M \to M$ is compatible with the twist map like this:

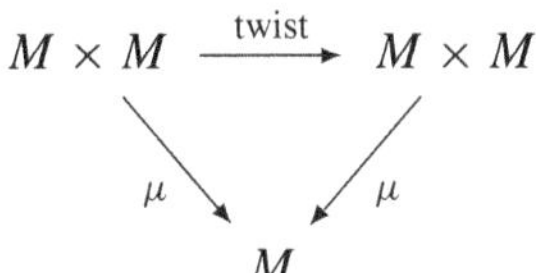

Examples

3.1.10 The natural numbers $\mathbb{N}$ with composition $+$ and neutral element 0 is the most important example of a monoid. We know well that this addition is associative and that 0 is neutral element for it. Note that $\mathbb{N}$ is commutative.

3.1.11 Groups are a special kind of monoid, characterised by the property that every element has an inverse.

3.1.12 Free monoids. Let S be a set (finite, say) whose elements we call letters. Consider the set of all words we can make out of these letters, including the empty word, i.e. the set of finite ordered sequences of elements of S. This set is a monoid when equipped with the operation of concatenation of words – clearly this is an associative operation. The neutral element is the empty word. This monoid is called the *free monoid* on S. ('Free' in this context means 'no relations', as opposed to 'free' in free software.) We consider S as a subset of the free monoid it generates, identifying the letters with 1-letter words...

Note that this monoid is not commutative (unless S is a singleton set), because the two words ab and ba are not equal.

As a variation we could just declare them to be equal, quotienting by the relations $ab = ba$, $a, b \in S$; this gives the notion of *free commutative monoid.* (So 'free commutative' means 'no relations other than the commutativity relations'.)

3.1.13 The natural numbers $\mathbb{N}$ *is the free monoid on a single generator.* To be concrete, take a single-element set $S = \{x\}$, then the free monoid generated by S is the set $\{x^0, x^1, x^2, x^3, \ldots\}$ where we used the notation x^0 to denote the empty word. Note that this monoid is automatically commutative. Now we have a monoid homomorphism

$$\begin{aligned}(\mathbb{N}, +, 0) &\longrightarrow \{x^0, x^1, \ldots\}\\ n &\longmapsto x^n\end{aligned}$$

which is clearly an isomorphism of monoids. So every free monoid on a single generator is isomorphic to $\mathbb{N}$, and we will allow ourselves to say that $\mathbb{N}$ is *the* free monoid on one generator – its generator is 1, and every element can be written in a unique way as a finite sum of 1s.

3.1.14 The integers $\mathbb{Z}$ (under addition) is a monoid which is not free. Indeed, you need as least two generators, a positive and a negative number, say $+1$ and -1. But then there will always be a relation, in this case $(+1) + (-1) = 0$. So $(\mathbb{Z}, +, 0)$ is not free.

3.1.15 A universal property of $\mathbb{N}$**.** Every element x in a monoid $(M, \cdot, 1)$ generates a submonoid $\{x^0, x^1, x^2, \ldots\}$, which in turn is the image of a unique monoid homomorphism

$$\begin{aligned}\mathbb{N} &\longrightarrow M\\ 1 &\longmapsto x,\end{aligned}$$

as in the preceding paragraph. In this way we get a canonical bijection

$$\{\text{elements of } M\} \leftrightarrow \boldsymbol{Mon}(\mathbb{N}, M).$$

(Compare 3.1.2.)

In the paragraphs above, 'free' meant 'no relations'. But the deeper meaning of the word is to possess a universal property like this.

3.1.16 The trivial monoid. What is the free monoid on the empty set? It is the set of words you can make out of no letters. There is only one such word namely the empty word. So we get the singleton monoid $(\{\varnothing\}, \cdot, \varnothing)$, called the *trivial monoid*. This notation is a bit confusing – it is better to stick to multiplicative notation and call the empty word 1. Also, since this monoid has only a single element 1, we will denote the monoid 1, just as we do for the singleton in the category of sets. The trivial monoid enjoys two universal properties. It is an initial object in ***Mon***: this means that given any monoid M there is a unique monoid homomorphism $1 \to M$; indeed, a monoid homomorphism must respect the identity. And second, it is a terminal object: given any monoid M there is a unique monoid homomorphism $M \to 1$. (Note that in the category of sets, the singleton set 1 is terminal, but not initial!)

3.1.17 The multiplicative monoid of $\mathbb{N}$ should be mentioned here, although we will not really use it: clearly, $(\mathbb{N}, \cdot, 1)$ is a monoid. Here is an example of a monoid homomorphism:

$$\begin{aligned} (\mathbb{N}, +, 0) &\longrightarrow (\mathbb{N}, \cdot, 1) \\ n &\longmapsto 7^n \end{aligned}$$

(by the familiar rules $7^a \cdot 7^b = 7^{a+b}$, and $7^0 = 1$). Of course the number 7 was chosen arbitrarily – we might as well have chosen any other natural number.

3.1.18 Endomorphism monoids. Let $X \in \boldsymbol{Top}$ be your favourite topological space and consider the set $\mathrm{End}_{\boldsymbol{Top}}(X)$ of all continuous maps from X to itself. Since we can compose such maps

$$\begin{aligned} \mathrm{End}(X) \times \mathrm{End}(X) &\longrightarrow \mathrm{End}(X) \\ (f, g) &\longmapsto fg, \end{aligned}$$

composition is associative, and we have the identity map $\mathrm{id} : X \to X$, we see that $\mathrm{End}(X)$ is a monoid. (Note that this is usually not commutative.)

More generally, let $\boldsymbol{C}$ be any category and let X be an arbitrary object in $\boldsymbol{C}$: then the set $\boldsymbol{C}(X, X)$ is a monoid. This follows immediately from the definition of a category, namely associativity of composition of arrows and existence of identity arrows.

3.1.19 Monoids as categories with a single object. We can make a small variation on this theme and state that *a monoid is essentially the same as a category with only one object*. Precisely, given a category $\boldsymbol{C}$ with only one object X, we have seen that $\boldsymbol{C}(X, X)$ is a monoid, and this monoid encodes all the data of the category. Conversely, given a monoid M, construct a category $\boldsymbol{C}$ by taking a single object (an arbitrary symbol X) and define the arrows to be $\boldsymbol{C}(X, X) := M$, such that composition of arrows is given by multiplication of elements in M (then necessarily the neutral element of the monoid must become the identity arrow of X). There was a choice of object, but if we compare two one-object categories coming from the same monoid, but with different single object, we immediately see they are isomorphic. Here is the dictionary:

$$\begin{aligned} \text{monoid } M &\leftrightarrow \text{category } \boldsymbol{C} \text{ with only one object } X \\ M &\leftrightarrow \boldsymbol{C}(X, X) \\ \text{element } m \in M &\leftrightarrow \text{arrow } m : X \to X \\ \text{neutral element } e &\leftrightarrow \text{identity arrow } \mathrm{id} : X \to X \\ \text{multiplication of elements} &\leftrightarrow \text{composition of arrows} \\ \text{monoid homomorphism} &\leftrightarrow \text{functor between one-object categories} \end{aligned}$$

3.1.20 Construction of multiplication in $\mathbb{N}$**.** Suppose we did not know there was a multiplication on $\mathbb{N}$. Then we could construct it, exploiting the universal property of $(\mathbb{N}, +, 0)$. First note that since $(\mathbb{N}, +, 0)$ is an object in the category of monoids, we can consider the endomorphism monoid $\mathrm{End}_{\boldsymbol{Mon}}(\mathbb{N}) = \boldsymbol{Mon}(\mathbb{N}, \mathbb{N})$, as in 3.1.18.

Now the crucial observation is that we have a canonical bijection (cf. 3.1.15)

$$\begin{aligned} \mathbb{N} &\leftrightarrow \mathrm{End}_{\boldsymbol{Mon}}(\mathbb{N}) \\ n &\leftrightarrow \varphi_n := [1 \mapsto n]. \end{aligned}$$

Since $\varphi_n : \mathbb{N} \to \mathbb{N}$ is a monoid homomorphism, it then maps $2 \mapsto n + n$, $3 \mapsto n + n + n$, and so on. (To restate the argument: a monoid homomorphism on $\mathbb{N}$ is completely determined by its value on 1 since $\mathbb{N}$ is freely generated by 1.) Now $\mathrm{End}_{\boldsymbol{Mon}}(\mathbb{N})$ is itself a monoid (via composition of endomorphisms), so we can just copy that monoid structure back to $\mathbb{N}$ via the bijection, and there we have it, the new composition law on $\mathbb{N}$, called multiplication! To see that it works let us compute what 3 times 7 is. By definition we must compose $1 \mapsto 3$ with $1 \mapsto 7$. So what does the second do to the result of the first? well, since it is a monoid homomorphism, it takes $3 = 1 + 1 + 1$ to $7 + 7 + 7$. So the composition takes 1 to $7 + 7 + 7$, and by definition this is then 3 times 7.

Monoid actions and representations

3.1.21 Actions. . . A right action of a monoid M on a set S is a map (of sets)

$$\alpha : S \times M \to S$$

such that these two diagrams commute:

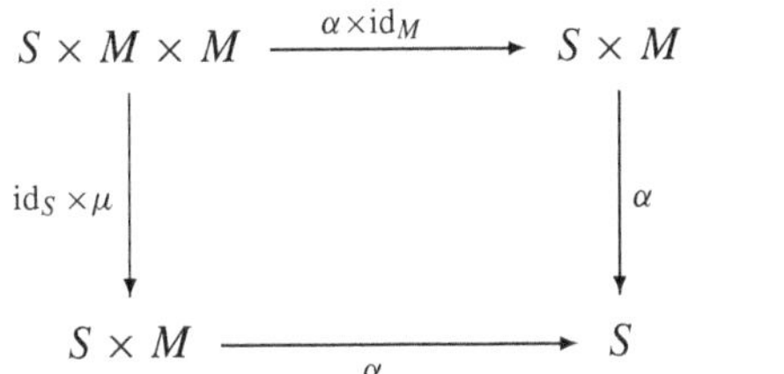

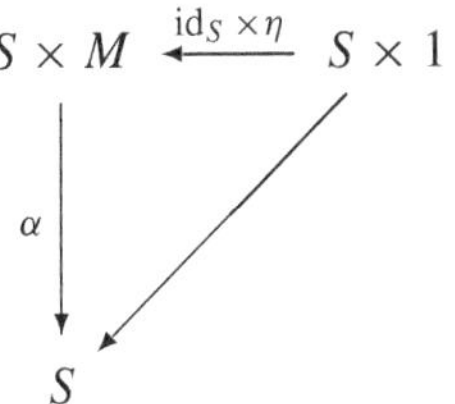

If we write the action like this:

$$\begin{aligned} S \times M &\xrightarrow{\alpha} S \\ (x, m) &\longmapsto x.m \end{aligned}$$

then we can write the axioms in terms of elements:

$$(x.m).m' = x.(m.m') \qquad \text{and} \qquad x.e = x.$$

3.1.22 ... in terms of endomorphisms cf. 3.1.18. Alternatively, this all amounts to giving a monoid homomorphism

$$M \to \mathrm{End}_{\boldsymbol{Set}}(S).$$

(Note that according to our left-to-right convention for composition, we must use right actions (not left actions) for this to work.)

3.1.23 Categorical viewpoint. Recall from 3.1.19 how the monoid M is interpreted as a category $\boldsymbol{M}$ with one single object. Similarly $\mathrm{End}_{\boldsymbol{Set}}(S)$ is regarded as the full subcategory of ***Set*** consisting of the single object S. In this setting we can now say that a monoid action of M is a functor $\boldsymbol{M} \to \boldsymbol{Set}$. The set on which M acts is then the image of the unique object of the category $\boldsymbol{M}$.

3.1.24 Linear representations. If we replace the category of sets with the category $\boldsymbol{Vect}_{\Bbbk}$ of vector spaces over $\Bbbk$ and $\Bbbk$-linear maps, then we obtain the notion of a linear representation of a monoid M on a vector space V: it is a monoid homomorphism $M \to \mathrm{End}_{\boldsymbol{Vect}_{\Bbbk}}(V)$. Again we can say more generally (varying V) that a linear representation of a monoid M is a functor $\boldsymbol{M} \to \boldsymbol{Vect}_{\Bbbk}$.

Exercises

1. Put $\mathbb{N}_+ = \mathbb{N} \smallsetminus \{0\}$. Show that $(\mathbb{N}_+, \cdot, 1)$ is the free commutative monoid generated by the set of all prime numbers. (Hint: unique factorisation.) If you want to include 0, you also need 0 among the generators, and then there are also relations, namely the relations $0 \cdot p = 0$ for each prime p.

2. Show that a bijective monoid homomorphism is an isomorphism (and vice versa).
3. Define a *semi-monoid* to be a set equipped with an associative composition law. (So it is like a monoid, but without the neutral element requirement.) Show that if a semi-monoid happens to have an element that satisfies the unit axiom, then this element is uniquely determined (i.e. a semi-monoid can have at most one neutral element). (Thus it makes sense to take a semi-monoid and ask whether it happens to be a monoid, and in this sense, 'having neutral element' is rather a property than a structure: one can regard the set of all monoids as a subset of the set of all semi-monoids.)
4. A *semi-monoid homomorphism* between two semi-monoids is a map compatible with the composition law, just like in the definition of monoid homomorphism. (i) Show that a semi-monoid homomorphism sends idempotents to idempotents. (Recall that an element x is an *idempotent* if $xx = x$.) Suppose two semi-monoids M and N happen to posses a neutral element (so they are actually monoids, cf. the previous exercise). Then every monoid homomorphism between them is clearly also a semi-monoid homomorphism. (ii) Give an example to show that the converse is not true: construct a semi-monoid homomorphism $M \to N$ that does not preserve the neutral elements, and thus is not a monoid homomorphism. (Hint: you need to take N to be a monoid with a nontrivial idempotent, cf. (i).)
   (The existence of such homomorphisms shows that while **Mon** can be considered a subcategory in the category of semi-monoids, it is not a full subcategory, cf. A.2.8.)
5. Let $(M', \ , e')$ and $(M, \ , e)$ be monoids, but consider them only to be semi-monoids by forgetting that they happen to posses a neutral element. Let $\phi : M \to N$ be a semi-monoid homomorphism (i.e. not required to preserve the neutral element). Show that if e is in the image of ϕ then ϕ is a monoid homomorphism (i.e. $e'\phi = e$).
6. Consider the special case of 3.1.22 where S is the underlying set of M (i.e. M acts on itself). Show that $M \to \mathrm{End}_{\mathbf{Set}}(M)$ is injective.

3.2 Monoidal categories

3.2.1 Note. Monoidal categories are also called *tensor categories* by many authors (e.g. Kassel [29]), because the category of vector spaces and tensor products is in many respects the key example (see 3.2.28).

3.2.2 Motivation. The diagrammatic definition of monoid given on page 140 relied on the following notions: *set*, *set map*, and *cartesian product of sets*.

Now instead of sets we can use any category where the notion of cartesian product makes sense. For example we can use *topological spaces*, *continuous maps*, and *cartesian products of topological spaces*, and then simply repeat the definitions of 3.1 in this setting. This leads to the notion of topological monoid: precisely, a topological monoid is a topological space X equipped with two continuous maps $\mu : X \times X \to X$ and $\eta : 1 \to X$, satisfying the axioms expressed by the commutative diagrams in 3.1.4. (So what is 1 here? it is the singleton topological space – or if you want: the empty cartesian product of topological spaces.)

Now a place where we would really like to do this is in the category of vector spaces. Clearly we could just repeat the constructions using *vector spaces*, *linear maps*, and *products of vector spaces*; the crucial structure would then be a linear map $V \times V \to V$. However, experience tells us that such linear maps are not nearly as interesting and useful as *bilinear maps* – maps which are linear in each variable – see Chapter 2. To capture this we should use the tensor product $\otimes$ instead of the cartesian product.

In other contexts we would like to use disjoint sums $\coprod$ instead – for example, disjoint sums played an important rôle in Chapter 1.

So we would like to extend the notion of monoid to refer to other 'binary' structures than just cartesian products. Such a structure should of course have properties similar to those properties of cartesian products that we used in the definition of monoid (listed in the beginning of this chapter). The important property is that in fact there is nothing too special about 'products' with two factors – it is mostly a generating concept: there are induced n-ary 'products' for all n, and then the nullary 'product' defines a 'neutral object' for the other 'products'. Structures of this sort are called *monoidal structures*, and categories equipped with such a structure are called *monoidal categories*.

So our first task is to define what it means to have this monoidal structure for a *category* – this will then serve as background for defining monoid structure on the *objects* of the category.

3.2.3 Cartesian products of categories. The important thing to note for this to make sense is that we have the notion of cartesian product of categories. For each pair of categories $\boldsymbol{C}$ and $\boldsymbol{D}$, there is a category $\boldsymbol{C} \times \boldsymbol{D}$ defined as follows: its objects are pairs (X, Y), where X is an object in $\boldsymbol{C}$ and Y is an object in $\boldsymbol{D}$ (in other words, the object set of $\boldsymbol{C} \times \boldsymbol{D}$ is the cartesian product of the object sets of $\boldsymbol{C}$ and $\boldsymbol{D}$). The set of arrows from (X, Y) to (X', Y') is the cartesian product $\boldsymbol{C}(X, X') \times \boldsymbol{D}(Y, Y')$. The empty product category (product of zero factors) is denoted $\mathbf{1}$. It is the category with only a single object, and only a single arrow (the identity arrow of the object).

Just as for sets, we have natural identifications $(\boldsymbol{C} \times \boldsymbol{D}) \times \boldsymbol{E} = \boldsymbol{C} \times (\boldsymbol{D} \times \boldsymbol{E})$, and $\mathbf{1} \times \boldsymbol{C} = \boldsymbol{C} = \boldsymbol{C} \times \mathbf{1}$, and thus we do in fact have n-ary products of categories for all $n \in \mathbb{N}$. In particular, let $\boldsymbol{C}^n$ denote the n-fold product of $\boldsymbol{C}$ with itself.

From now on we take for granted (and in fact rely completely on) the notions of functors and natural transformations. The reader unfamiliar with these notions should consult the Appendix.

Definition of monoidal categories

Our notion of monoidal categories will be the notion of *strict* monoidal categories, and the adjective will appear in the definition for the sake of precision. There is a weaker notion of monoidal category which is actually the 'correct' one, but which is somewhat more complicated. Below (on page 154), we will explain the difference and justify the abuse we commit when pretending that all monoidal categories are strict.

3.2.4 Monoidal categories. A *(strict) monoidal category* is a category $\mathbf{V}$ together with two *functors*

$$\mu : \mathbf{V} \times \mathbf{V} \to \mathbf{V}, \qquad \eta : \mathbf{1} \to \mathbf{V}$$

satisfying the associativity axiom and the neutral object axiom. Precisely we require that these three diagrams commute:

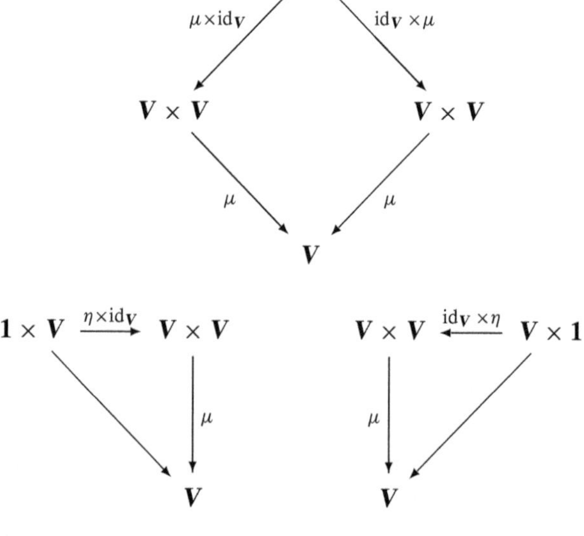

The symbols $\mathrm{id}_{\mathbf{V}}$ stand for the identity functor $\mathbf{V} \to \mathbf{V}$, and the diagonal functors without label are the projections, which are canonical identifications.

Let us stress that the μ and η are functors. This means that they operate on both objects and arrows, as we now spell out in detail.

3.2.5 Details concerning the functor $\mu : \mathbf{V} \times \mathbf{V} \to \mathbf{V}$. Since we are going to use monoidal categories as background for working with monoids (with μ as replacement of cartesian product), it is practical to adopt infix notation for μ, so we write

$$\begin{aligned} \mathbf{V} \times \mathbf{V} &\xrightarrow{\mu} \mathbf{V} \\ (X, Y) &\longmapsto X \square Y \\ (f, g) &\longmapsto f \square g. \end{aligned}$$

(In the applications, $\square$ will be $\times$, $\otimes$, $\coprod$, or the like.) So to each pair of objects X, Y, a new object $X \square Y$ is associated, and to each pair of arrows $X \xrightarrow{f} X'$, $Y \xrightarrow{g} Y'$ a new arrow $X \square Y \xrightarrow{f \square g} X' \square Y'$. The fact that $\square$ is a functor means that compositions and identity arrows are respected. In detail, given compositions $X \xrightarrow{f} X' \xrightarrow{f'} X''$, $Y \xrightarrow{g} Y' \xrightarrow{g'} Y''$ then we have

$$ff' \square gg' = (f \square g)(f' \square g')$$

this is an equality of arrows $X \square Y \to X'' \square Y''$. Concerning identity arrows: given $\mathrm{id}_X : X \to X$ and $\mathrm{id}_Y : Y \to Y$ we have

$$\mathrm{id}_X \square \mathrm{id}_Y = \mathrm{id}_{X \square Y} .$$

3.2.6 The monoidal operator as paralleling. We draw an arrow $f : X \to X'$ like this:

$$\underset{X}{\bullet} \xrightarrow{\quad f \quad} \underset{X'}{\bullet}$$

and a composition $X \xrightarrow{f} X' \xrightarrow{f'} X''$ like this:

$$\underset{X}{\bullet} \xrightarrow{\quad f \quad} \underset{X'}{\bullet} \xrightarrow{\quad f' \quad} \underset{X''}{\bullet}$$

If we think of arrows in a category as *processes* then composition of two arrows means: let the first process terminate, and then apply the second process to the result. The identity arrow means 'do nothing'!

Now the monoidal operator expresses the concept of parallel processes. We draw $X \square Y$ like this:

$$\begin{array}{l} Y \bullet \\ X \bullet \end{array}$$

with the second 'factor' on top of the first, and draw the arrow $f \Box g$ like this:

$$\begin{array}{ccc} & g & \\ Y \bullet & \longrightarrow & \bullet Y' \\ X \bullet & \longrightarrow & \bullet X' \\ & f & \end{array}$$

this means that we run the two processes in parallel. And finally the functor condition states that it makes no difference whether we take the serial connection of two parallel connections or the parallel connection of the two serial connections; in any case the result is this process:

$$\begin{array}{ccccc} & g & & g' & \\ Y \bullet & \longrightarrow & \bullet & \longrightarrow & \bullet Y'' \\ X \bullet & \longrightarrow & \bullet & \longrightarrow & \bullet X'' \\ & f & & f' & \end{array}$$

In particular this gives a useful possibility of performing a 'complicated parallel processing situation' $f \Box g$ (the middle picture), by letting g wait for f (left-hand picture) or letting f wait for g (right-hand picture):

$$\begin{array}{ccccc} & \mathrm{id}_Y & & g & \\ Y \bullet & \text{---} & \bullet & \longrightarrow & \bullet Y' \\ X \bullet & \longrightarrow & \bullet & \text{---} & \bullet X' \\ & f & & \mathrm{id}_{X'} & \end{array} = \begin{array}{ccc} & g & \\ Y \bullet & \longrightarrow & \bullet Y' \\ X \bullet & \longrightarrow & \bullet X' \\ & f & \end{array} = \begin{array}{ccccc} & g & & \mathrm{id}_{Y'} & \\ Y \bullet & \longrightarrow & \bullet & \text{---} & \bullet Y' \\ X \bullet & \text{---} & \bullet & \longrightarrow & \bullet X' \\ & \mathrm{id}_X & & f & \end{array}$$

In these pictures, the arrow head was omitted on the identity arrows (waiting processes), to emphasise that nothing happens in this thread... And henceforth we will suppress the small arrow heads altogether, just to avoid clutter, and recognise the direction of the arrow by the convention that *arrows go from the left to the right* unless otherwise specified.

3.2.7 Details concerning $\eta : \mathbf{1} \to \mathbf{V}$. Let I denote the object which is the image of $\eta : \mathbf{1} \to \mathbf{V}$. Then the statement of the two triangular diagrams can be formulated like this:

$$I \Box X = X = X \Box I, \qquad \mathrm{id}_I \Box f = f = f \Box \mathrm{id}_I \tag{3.2.8}$$

for every object X, and for every arrow f.

We refer to a monoidal category by specifying the triple $(\mathbf{V}, \Box, I)$.

3.2.9 n-ary products. In view of associativity, the functor $\mu : \mathbf{V} \times \mathbf{V} \to \mathbf{V}$ induces functors $\mu^{(n)} : \mathbf{V}^n \to \mathbf{V}$ for all $n \geq 2$. For completeness we also let

$\mu^{(1)} : \mathbf{V}^1 \to \mathbf{V}$ be the identity functor $\mathrm{id}_{\mathbf{V}} : \mathbf{V} \to \mathbf{V}$, and let $\mu^{(0)} : \mathbf{V}^0 \to \mathbf{V}$ be the neutral object functor $\eta : \mathbf{1} \to \mathbf{V}$. (Compare 3.1.5.)

3.2.10 Monoidal functors. A *(strict) monoidal functor* between two (strict) monoidal categories $(\mathbf{V}, \Box, I)$ and $(\mathbf{V}', \Box', I')$ is a functor $F : \mathbf{V} \to \mathbf{V}'$ that commutes with all the structure. Precisely, these two diagrams are required to commute:

$$\begin{array}{ccc} \mathbf{V} \times \mathbf{V} & \xrightarrow{F \times F} & \mathbf{V}' \times \mathbf{V}' \\ {\scriptstyle \mu}\downarrow & & \downarrow{\scriptstyle \mu'} \\ \mathbf{V} & \xrightarrow[F]{} & \mathbf{V}' \end{array} \qquad \begin{array}{ccc} \mathbf{V} & \xrightarrow{F} & \mathbf{V}' \\ {\scriptstyle \eta}\uparrow & & \uparrow{\scriptstyle \eta'} \\ \mathbf{1} & = & \mathbf{1} \end{array}$$

So in terms of objects we have

$$(XF)\Box'(YF) = (X\Box Y)F \quad \text{and} \quad IF = I',$$

and in terms of arrows we have

$$(fF)\Box'(gF) = (f\Box g)F.$$

In terms of n-ary products we can write the requirements uniformly as

$$\begin{array}{ccc} \mathbf{V}^n & \xrightarrow{F^n} & \mathbf{V}^n \\ {\scriptstyle \mu^{(n)}}\downarrow & & \downarrow{\scriptstyle \mu^{(n)}} \\ \mathbf{V} & \xrightarrow[F]{} & \mathbf{V} \end{array} \qquad \text{for all } n \geq 0.$$

3.2.11 The category of monoidal categories. One easily checks that the composition of two monoidal functors is again monoidal, and that identity functors are monoidal, so all together there is a category denoted ***MonCat*** whose objects are the monoidal categories and whose arrows are the monoidal functors. A monoidal functor is called an isomorphism of monoidal categories if there exists a two-sided inverse in ***MonCat***.

3.2.12 Examples of monoidal categories will be given in the subsection starting on page 157. Unfortunately, the most important ones (including $(\textbf{\textit{Set}}, \times, 1)$, $(\textbf{\textit{Set}}, \coprod, \emptyset)$, $(\textbf{\textit{Vect}}_{\Bbbk}, \otimes, \Bbbk)$, $(\textbf{\textit{2Cob}}, \coprod, \emptyset)$) are not really *strict* monoidal categories, so in order not to cheat the reader (and to save her from an unpleasant surprise when she picks up a random book on monoidal categories) we should spend a short while with not-necessarily-strict monoidal categories.

Nonstrict monoidal categories

See Mac Lane [34] or Kassel [29] for more details.

There is a weaker notion of monoidal categories, usually simply called *monoidal categories*, where the axioms only hold up to coherent isomorphisms instead of holding strictly. In this subsection we briefly give the definition, and explain why there is no harm in pretending that all monoidal categories are strict. (If the reader finds this confusing he should rather skip this subsection – it will not really be used elsewhere in the text.)

The axioms for a strict monoidal category state that certain diagrams of functors commute (cf. 3.2.4); each of the three diagrams expresses the equality of two functors. Now instead of having equality we just require an invertible natural transformation between the two functors; this natural transformation is part of the structure and must be specified. These natural transformations must satisfy certain *coherence constraints* which guarantee that treating the corresponding isomorphisms (the components of the natural transformations) as if they were equalities will not lead to contradictions.

3.2.13 Weak associativity. The associativity axiom for a strict monoidal category $(\mathbf{V}, \mu, \eta)$ states the equality of the two functors

$$\mathbf{V} \times \mathbf{V} \times \mathbf{V} \underset{(\mathrm{id}_{\mathbf{V}} \times \mu)\mu}{\overset{(\mu \times \mathrm{id}_{\mathbf{V}})\mu}{\rightrightarrows}} \mathbf{V}.$$

Now instead of equality we require just an invertible natural transformation

$$\alpha : (\mu \times \mathrm{id}_{\mathbf{V}})\mu \Rightarrow (\mathrm{id}_{\mathbf{V}} \times \mu)\mu,$$

called the *associator*. It is suggestive to draw a natural transformation as a 2-*cell* like this:

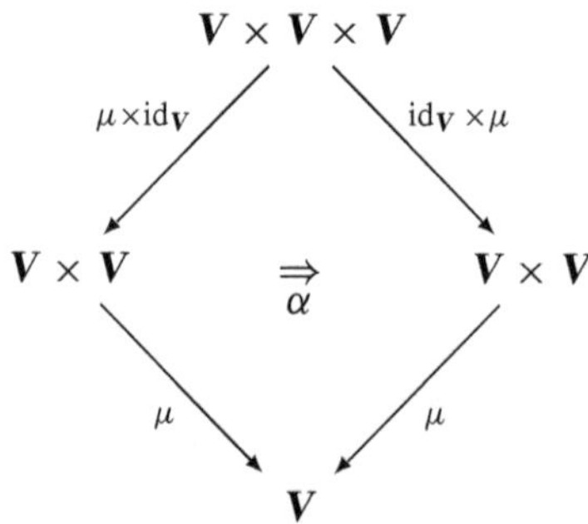

This means that the diagram is *not* commutative, but rather that the natural transformation α compares the two possible ways of going through the diagram. In concrete terms: for every triple of objects X, Y, Z there is an isomorphism $\alpha_{X,Y,Z} : (X\Box Y)\Box Z \xrightarrow{\sim} X\Box(Y\Box Z)$. Naturality means that these isomorphisms are compatible with all arrows (see Appendix).

Here comes the coherence constraint. For every quadruple of objects A, B, C, D, if we start with the 'product' $((A\Box B)\Box C)\Box D$ then there are two ways we can use the associator to shuffle all the parentheses over to the right, and the coherence constraint on α states that it does not matter which way we choose, the result is not the same. Precisely we require the following commutative pentagon diagram:

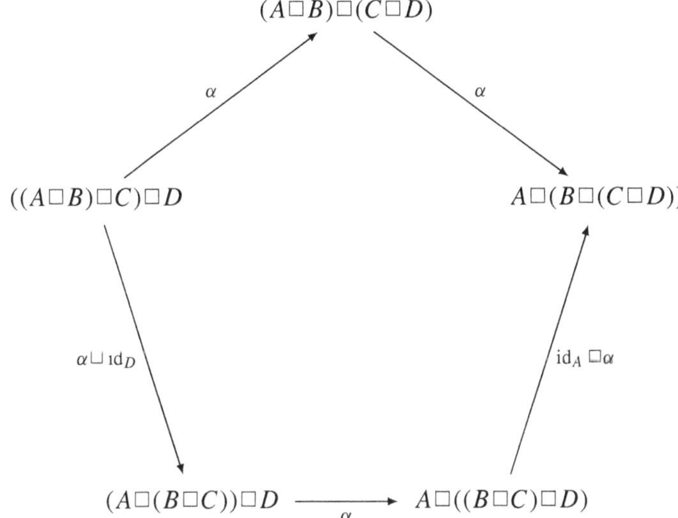

(The α here are really the components of α corresponding to the object in question. That is, they ought to be indexed (like $\alpha_{A\Box B,C,D}$ for the upper left-hand arrow).)

3.2.14 Weak unit axioms. Instead of having strict triangles (as in 3.2.4), we only have invertible 2-cells (denoted λ and ρ)

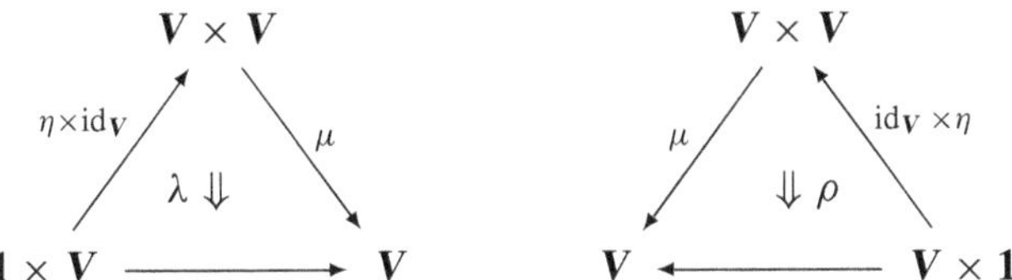

These natural equivalences amount to having for each object A natural isomorphisms

$$\lambda_A : I\Box A \xrightarrow{\sim} A, \qquad \rho_A : A\Box I \xrightarrow{\sim} A.$$

These natural equivalences are also subject to a coherence constraint, namely the commutativity of this diagram (for all A, B):

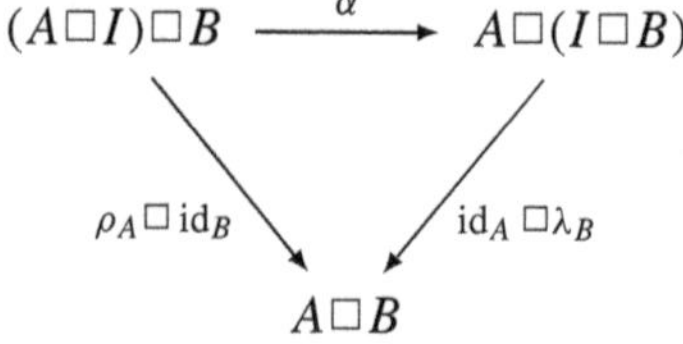

3.2.15 The definition. A *monoidal category* is a sextuple $(\boldsymbol{V}, \mu, \eta, \alpha, \lambda, \rho)$ where $\boldsymbol{V}$ is a category, $\mu : \boldsymbol{V} \times \boldsymbol{V} \to \boldsymbol{V}$ and $\eta : \mathbf{1} \to \boldsymbol{V}$ are functors, and α, λ, ρ are invertible natural transformation as above satisfying the coherence constraints.

3.2.16 Strict monoidal categories. A monoidal category in the sense of this definition is called *strict* if α, λ, ρ are all the identity natural transformations.

Note that this weakening principle has no analogue for monoids (in ***Set***) since there is no such thing as a 'natural transformation' between arrows in ***Set*** (functions).

3.2.17 Monoidal functors come in different flavours, depending on the level of strictness: *strict* monoidal functors are functors that respect all the structure $(\boldsymbol{V}, \mu, \eta, \alpha, \lambda, \rho)$. (The functors we use are always considered strict.) *Strong* monoidal functors are not required to preserve the structure on the nose: instead of the identity maps which express the structure preservation there should be invertible comparison morphisms (e.g. $XF\square' YF \xrightarrow{\sim} (X\square Y)F$). These should then be specified and are part of the data of a monoidal functor. For the precise definition, see Mac Lane [34], Chapter XI, Section 2. Finally, for *lax* monoidal functors (sometimes simply called monoidal functors), these comparison morphisms are no longer required to be invertible. (In any case the comparison morphisms must satisfy certain coherence constraints.) Note that in the strict context 'being monoidal' is a *property* of a functor: it makes sense to take a functor (between monoidal categories) and ask whether it is monoidal. In the nonstrict context 'being monoidal' is a *structure* which must be specified.

Next, there is a notion of a strong monoidal functor being a *monoidal equivalence*; it means that there exists a strong monoidal functor in the other direction such that their two compositions are isomorphic to the identity functors (under a monoidal natural transformation – a notion which must also be defined properly – see 3.2.49 for the strict version of this). We mention these notions only to be able to state the

3.2.18 Strictification Theorem. (See Mac Lane [34], Chapter XI, Section 3.) *Every monoidal category is monoidally equivalent to a strict monoidal category.* □

This theorem is essentially equivalent to

3.2.19 Mac Lane's Coherence Theorem. (See Mac Lane [34], Chapter VII, Section 2.) *Let* $(\mathbf{V}, \square, I, \alpha, \lambda, \rho)$ *be a monoidal category. Every diagram that can be built out of the components of* α, λ, ρ, *and identity maps, using composition and monoidal operations, commutes.* □

In other words, the coherence constraints expressed by the commutativity of the pentagon diagrams and the triangle diagrams imply general coherence (expressed by the commutativity of all other diagrams which represent different ways of shuffling parentheses (or deleting copies of I)).

In practice this means that there is no harm done in pretending that all the comparison isomorphisms are actually identity maps, and thus that the monoidal category is strict. This is what we do throughout.

Examples of monoidal categories and functors

3.2.20 Discrete monoidal categories. From the definition we see that monoidal categories are to categories as monoids are to sets. Let us make that remark functorial. A category whose only arrows are the identity arrows is called *discrete*. Thus a discrete category is specified completely by specifying its objects. Conversely, every set S can be considered a discrete category $\boldsymbol{S}$: just take the objects of $\boldsymbol{S}$ to be the elements of S, and take no arrows other than the identity arrows. Under this correspondence, functions between sets translate into functors between (discrete) categories. All told, there is a functor

$$\boldsymbol{Set} \hookrightarrow \boldsymbol{Cat}$$

whose image consists of (the) discrete categories.

If now S is a monoid then the structure maps $\mu : S \times S \to S$ and $\eta : 1 \to S$ translate into structure functors $\mu : \boldsymbol{S} \times \boldsymbol{S} \to \boldsymbol{S}$ and $\eta : \mathbf{1} \to \boldsymbol{S}$, and the monoid axioms that hold for these maps translate into the axioms for a monoidal category: so $\boldsymbol{S}$ is a (discrete) monoidal category.

3.2.21 The category of sets, *equipped with cartesian product and singleton set is a monoidal category.* To each pair of sets (S, S') we can associate the set $S \times S'$:

$$\begin{aligned} \boldsymbol{Set} \times \boldsymbol{Set} &\longrightarrow \boldsymbol{Set} \\ (S, S') &\longmapsto S \times S'. \end{aligned}$$

So $\Box$ is simply the cartesian product $\times$ itself! Note however that the two $\times$ symbols that occur in the definition are not the same: ***Set*** $\times$ ***Set*** denotes the cartesian product of two categories, while $S \times S'$ denotes the cartesian product of two sets.

The singleton set 1 serves as neutral element for this operation. It is the image of the functor

$$\begin{aligned} \mathbf{1} &\longrightarrow \textbf{\textit{Set}} \\ * &\longmapsto 1. \end{aligned}$$

More generally, this works for

3.2.22 Any category that admits products. *Let* $\mathbf{C}$ *be a category that admits products – in particular let* 1 *denote the empty product in* $\mathbf{C}$*, which is 'the' terminal object. Then* $(\mathbf{C}, \times, 1)$ *is a monoidal category.*

Let us have a short look at how the universal property guarantees that the axioms are satisfied – we concentrate on the associativity axiom. We want to show that for each triple of objects A, B, C we have

$$(A \times B) \times C \simeq A \times (B \times C).$$

First let us recall that $(A \times B) \times C$ is characterised (up to unique isomorphism) by the universal property

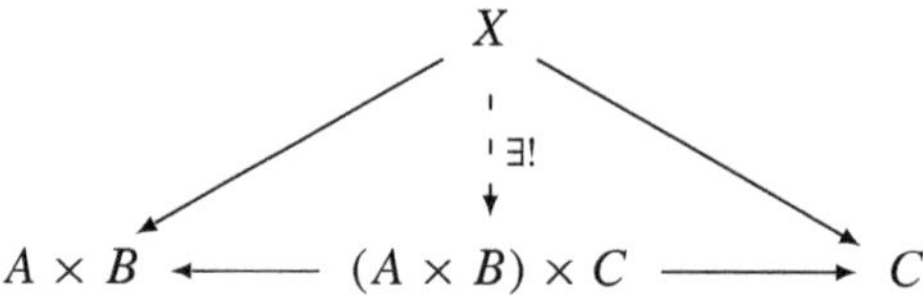

We want to put $A \times (B \times C)$ in the place of X. Note that we have projection maps

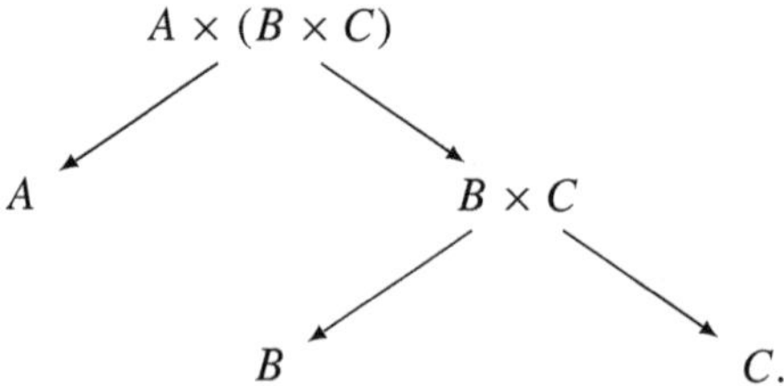

Since we have maps from $A \times (B \times C)$ to both A and B, the universal property of $A \times B$ gives us a canonical map $A \times (B \times C) \rightarrow A \times B$. Now we already had a map to C, so that is all we need to put $A \times (B \times C)$ in the place of X in the diagram. Now the universal property of $(A \times B) \times C$ gives us a map from $A \times (B \times C)$. Repeating all the arguments with the parentheses moved to the other side gives a map in the other direction, and it is not difficult to see that these two maps are in fact inverses to each other, so we have an isomorphism. Now the fact that these isomorphisms (one pair for each triple of objects) are all constructed by universal properties guarantees that they are coherent, so we have a monoidal category. To be precise it is not a strict monoidal category, but we will nevertheless regard it as such.

So ($\boldsymbol{Set}$, $\times$, 1) is a special case of this. Two other special cases are worth mentioning explicitly at this point.

3.2.23 The category of all (small) categories, ($\boldsymbol{Cat}$, $\times$, 1), *is a monoidal category.* As an example of a monoidal functor we have the inclusion

$$(\boldsymbol{Set}, \times, 1) \hookrightarrow (\boldsymbol{Cat}, \times, \mathbf{1})$$

described in 3.2.20.

3.2.24 The category of monoids ($\boldsymbol{Mon}$, $\times$, 1) *is a monoidal category.* (The product of two monoids was defined in 3.1.8 – its multiplication is just component-wise multiplication.) This example will be subsumed by Example 3.5.5 where more details can be found.

3.2.25 Any category that admits coproducts $\coprod$ *is a monoidal category.* The neutral object is then of course the initial object. This follows from the same reasoning as with products; the arrows constructed by the universal property go the other way around, but since anyway they are isomorphisms it makes no difference. Alternatively, this result can be obtained from 3.2.22 by considering the opposite category.

Concrete examples of coproduct monoidal structure:

3.2.26 The category of sets and disjoint union. For each pair of sets X, Y we can form their disjoint union $X \coprod Y$. We noted in 1.3.24 that this composition is associative (modulo natural identifications). Clearly the empty set $\emptyset$ is the neutral object. The variation 'finite sets and disjoint union' will be studied in more detail in Section 3.4.

3.2.27 The category of topological spaces, and disjoint union. As we saw in 1.3.25.

3.2.28 The category of vector spaces, and tensor product. Given two $\Bbbk$-vector spaces U, V, then we can form their tensor product $U \otimes V$. For three vector spaces U, V, W we have $(U \otimes V) \otimes W = U \otimes (V \otimes W)$. The ground field $\Bbbk$ itself is the neutral object.

Note that in reality, we only have *isomorphisms* $(U \otimes V) \otimes W \simeq U \otimes (V \otimes W)$, but these isomorphisms are natural and unique (when required to be compatible with the structure maps of the tensor products).

(There is no point in merely observing that the spaces are isomorphic (assuming that the spaces U, V, W are finite dimensional, clearly the two versions of the triple tensor product have the same dimension and are therefore isomorphic). The crux is really that we make the identification along the *correct* isomorphisms. Choosing other isomorphisms would lead to contradictions. . .)

3.2.29 R-modules and groups. For the same reason, if R is a commutative ring, the category $(\boldsymbol{Mod}_R, \otimes_R, R)$ of R-modules with the tensor product is monoidal. In particular, the category of abelian groups $(\boldsymbol{Ab}, \otimes_{\mathbb{Z}}, \mathbb{Z})$ is a monoidal category.

3.2.30 Another monoidal structure on $\boldsymbol{Vect}_{\Bbbk}$. The triple $(\boldsymbol{Vect}_{\Bbbk}, \oplus, \mathbf{0})$ is a monoidal category. (This is a special case of 3.2.22 and also 3.2.25.)

3.2.31 The category of Frobenius algebras (with tensor products). We saw in 2.4.8 that the tensor product of two Frobenius algebras is again a Frobenius algebra, and that the trivial Frobenius algebra $(\Bbbk, \mathrm{id}_{\Bbbk})$ is the neutral object for this operation. So $(\boldsymbol{FA}_{\Bbbk}, \otimes, \Bbbk)$ is a monoidal category.

3.2.32 The category of commutative Frobenius algebras (with tensor products). The tensor product of two commutative Frobenius algebras is again commutative, and clearly the trivial Frobenius algebra $(\Bbbk, \mathrm{id}_{\Bbbk})$ is commutative. So $(\boldsymbol{cFA}_{\Bbbk}, \otimes, \Bbbk)$ is a monoidal category (with a canonical monoidal embedding into $(\boldsymbol{FA}_{\Bbbk}, \otimes, \Bbbk)$).

3.2.33 The category of n-dimensional cobordisms $(\boldsymbol{nCob}, \coprod, \emptyset)$ is a monoidal category. This was explained in Chapter 1, page 48.

Symmetric monoidal categories

We are going to use monoidal categories as the context for defining monoids (cf. the introductory discussion on page 148). Now in order to be able to talk about *commutative* monoids, we need a notion of *twist map* in the category, generalising the twist map we know in the category of sets. Intuitively this new twist map will be 'interchange of factors'. When such a twist map is specified with properties similar to those of the twist map in **Set** then we have a *symmetric monoidal category*. (Specifying twist maps with slightly weaker properties leads to the notion of braided monoidal categories, which is not really used in this text (except for a brief mention in the exercises), but which is important in knot theory (see Kassel [29]).)

3.2.34 Definition. A (strict) monoidal category $(\mathbf{V}, \square, I)$ is called a *symmetric monoidal category* if for each pair of objects X, Y there is given a *twist map*

$$\tau_{X,Y} : X \square Y \to Y \square X$$

subject to the following three axioms (to be explained below).

(i) The maps are natural.
(ii) For every triple of objects X, Y, Z, these two diagrams commute:

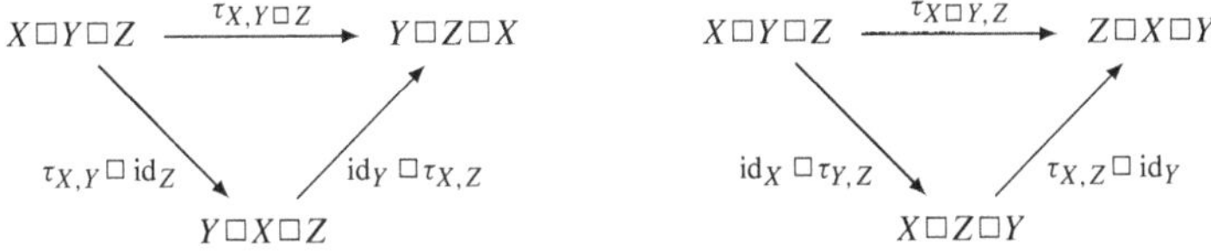

(iii) We have $\tau_{X,Y}\tau_{Y,X} = \mathrm{id}_{X \square Y}$.

3.2.35 The twist map is pictured like this:

Y X
X Y

Axiom (iii) should be compared to the first relation for the symmetric group (1.4.2). Its graphical representation is

Y Y = Y Y
X X X X

The collection of twist maps (one for each pair of objects) is a *structure* that must be specified (not a property of a given monoidal category), so a symmetric monoidal category is a quadruple $(\mathbf{V}, \square, I, \tau)$. (However, in many important cases there is no choice: when the monoidal structure is given by product or coproduct, the axioms are so strong that there is only one possible symmetric structure, cf. 3.2.42. We will see in Example 3.2.46 that the category of graded vector spaces admits more than one symmetry. . .)

3.2.36 The naturality condition (i) means that for every arrow in $\mathbf{V} \times \mathbf{V}$ (i.e. for every pair of arrows $f : X \to X'$ and $g : Y \to Y'$), the diagram

$$\begin{array}{ccc} X \square Y & \xrightarrow{\tau_{X,Y}} & Y \square X \\ {\scriptstyle f \square g}\downarrow & & \downarrow{\scriptstyle g \square f} \\ X' \square Y' & \xrightarrow[\tau_{X',Y'}]{} & Y' \square X' \end{array}$$

commutes. This is precisely to say that the collection of twist maps assemble into a natural transformation τ; it goes from

$$\mathbf{V}\times\mathbf{V}\xrightarrow{\mu}\mathbf{V} \qquad \text{to} \qquad \mathbf{V}\times\mathbf{V}\xrightarrow{\text{twist}}\mathbf{V}\times\mathbf{V}\xrightarrow{\mu}\mathbf{V}.$$

It is a 2-cell like this:

$$\begin{array}{ccc} \mathbf{V}\times\mathbf{V} & \xrightarrow{\text{twist}} & \mathbf{V}\times\mathbf{V} \\ {\scriptstyle\mu}\downarrow & \nearrow & \downarrow{\scriptstyle\mu} \\ \mathbf{V} & = & \mathbf{V} \end{array}$$

(In particular, one may note that the notion of twist map in a monoidal category depends on the existence of the canonical twist functor for categories (the twist map in ***Cat***).)

Let us look at an interesting case: f is the identity map on X and g is the twist map $Y\square Z\to Z\square Y$. That gives the commutative diagram

$$\begin{array}{ccc} X\square Y\square Z & \xrightarrow{\tau_{X,Y\square Z}} & Y\square Z\square X \\ {\scriptstyle \mathrm{id}_X\square\tau_{Y,Z}}\downarrow & & \downarrow{\scriptstyle \tau_{Y,Z}\square\mathrm{id}_X} \\ X\square Z\square Y & \xrightarrow[\tau_{X,Z\square Y}]{} & Z\square Y\square X \end{array}$$

It is useful to draw pictures – here is the graphical version of the commutative diagram:

Z Y X [diagram] X Y Z = Z Y X [diagram] X Y Z (3.2.37)

3.2.38 Symmetry. Condition (ii) says that the twist maps compose like permutations. In graphical guise the left-hand condition reads:

Z Y X [diagram] X Z Y = Z Y X [diagram] X Z Y

(That is, moving X past $Y\square Z$ can be achieved by the two-step operation: first moving X past Y and then past Z.) Now take this expression for [diagram] and plug it into each side of Equation 3.2.37 to get:

Z Y X [diagram] X Y Z = Z Y X [diagram] X Y Z (3.2.39)

in analogy with the relation for the symmetric groups, cf. 1.4.2: every element in a symmetric group can be written as a product of transpositions and the possible products satisfy this relation.

If we think of monoidal structure as 'n-ary products for all $n \in \mathbb{N}$' (which we can do in view of associativity), then a symmetry can be interpreted as a rule which to every n-tuple and to every permutation of its entries associates a 'higher twist map'. Now axiom (ii) says that composition of such maps behaves just like composition of permutations: the big collection of all these twist maps is generated by binary twist maps, just like the symmetric groups are generated by transpositions, and satisfy relation 3.2.39 which is analogous to the symmetric group relation.

(Note that Equation 3.2.39 could also be obtained from the right-hand part of condition (ii), and this also shows that that the two diagrams of condition (ii) in fact imply each other (modulo naturality).)

The axioms also imply a strong compatibility with the unit structure: twisting with the neutral object I has no effect, according to this lemma:

3.2.40 Lemma. *For every object X, these two diagrams commute:*

Proof. We will treat the left-hand triangle (the right-hand is analogous). Consider this tetrahedron (where we have suppressed the symbol $\square$ between objects to save space):

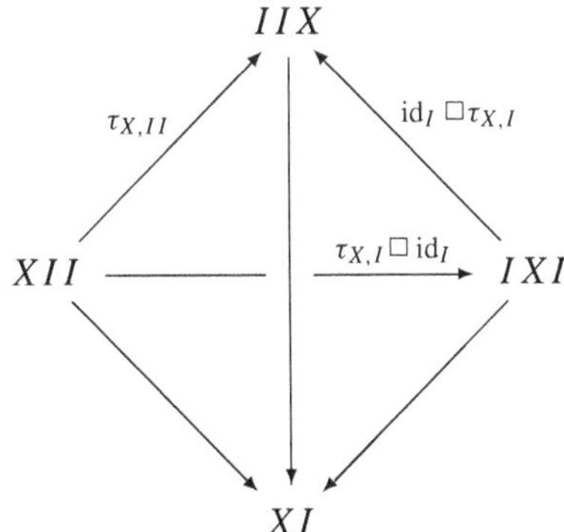

The base triangle is the triangle of the lemma (just with an extra factor $\square\, \mathrm{id}_I$). We will show the base triangle is commutative, by establishing the commutativity of the other triangles. The back triangle,

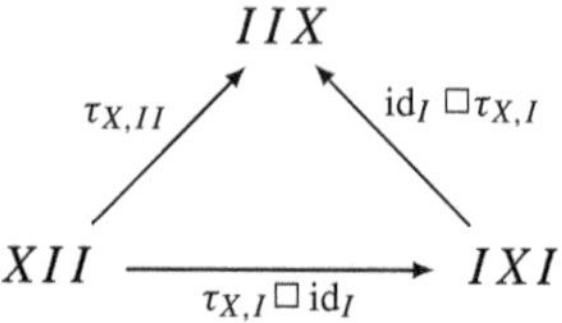

is just the left-hand triangle of axiom (ii), with $Y = Z = I$.

The front edge of the tetrahedron is actually the composite

$$IIX \longrightarrow IX \xrightarrow{(\tau_{X,I})^{-1}} XI.$$

Hence, the commutativity of the two front faces is equivalent to the two squares

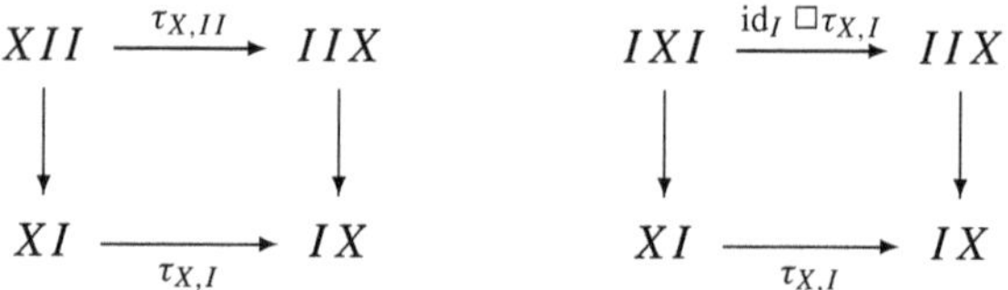

The left-hand square is just the naturality of τ with respect to id_X and $II \to I$. The right-hand square commutes since id_I acts neutrally on arrows. □

3.2.41 Example. Let $(\boldsymbol{M}, \Box, I)$ be a discrete monoidal category (i.e. a monoid) cf. 3.2.20. Since the only arrows are the identity arrows, the only possible symmetry structure there might exist on $\boldsymbol{M}$ is the one where each $\tau_{X,Y} : X \Box Y \to Y \Box X$ is the *identity*. Now the statement $X \Box Y = Y \Box X, \forall X, Y$ is something which is either true or false (it is a property not a structure), and if it is false then of course there can be no symmetry structure on $\boldsymbol{M}$. If it is true it amounts to saying that the corresponding monoid is commutative (cf. 3.1.9). So symmetric structure (in a monoidal category) generalises commutativity of monoids. But while it is a *structure* in monoidal categories, in monoids there is no choice for the symmetry: either it exists (commutative case) or it does not (noncommutative case), so it is a *property*.

3.2.42 Canonical symmetries. We saw in 3.2.22 that if $\boldsymbol{V}$ is a category with products, then $(\boldsymbol{V}, \times, 1)$ is a monoidal category (where 1 denotes the terminal object, the empty product). There is a canonical symmetric structure on $\boldsymbol{V}$ namely the one interchanging the two factors. (Exercise: write out the details of checking the axioms.) Now we claim that this is the only possible symmetry.

Indeed, given two objects A, B, consider the two naturality diagrams

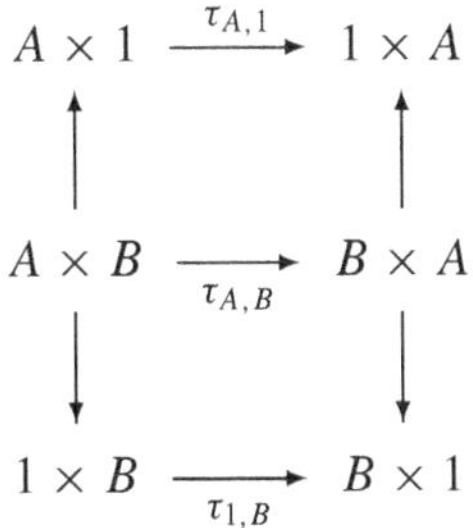

where the vertical maps are the projection maps (or more precisely perhaps: the unique map to 1 times the identity map). Now axiom (iii) for being a symmetry says that that twist on the neutral object is the identity (the upper and lower horizontal maps). So in conclusion the twist map $A \times B \to B \times A$ must also be a sort of identity map on each factor – but twisted, as in the definition of the canonical twist map described initially.

Clearly the same argument works to show that in a category whose monoidal structure comes from the coproduct, there is a unique symmetry structure.

3.2.43 Examples. All the monoidal categories we have encountered so far carry a canonical symmetry, even if the monoidal operation is neither product nor coproduct. In each case the twist map is the obvious interchange of 'factors'. Nonsymmetric monoidal categories will appear in Section 3.4: the monoidal category of finite ordered sets does not admit any symmetry. (It is precisely the ordering that prevents us from permuting anything. . .)

To be explicit with a couple of our favourite examples:

3.2.44 The category of n-cobordisms ($\boldsymbol{n}$**Cob**$, \coprod, \varnothing, T$), where the twist cobordism $T_{\Sigma,\Sigma'} : \Sigma \coprod \Sigma' \Rightarrow \Sigma' \coprod \Sigma$ is the cobordism induced from the twist diffeomorphism $\tau_{\Sigma,\Sigma'} : \Sigma \coprod \Sigma' \to \Sigma' \coprod \Sigma$ in the category of smooth manifolds.

3.2.45 The category of vector spaces (**Vect**$_{\Bbbk}, \otimes, \Bbbk$) carries a canonical symmetry σ which goes like this:

$$\begin{aligned} V \otimes W &\longrightarrow W \otimes V \\ v \otimes w &\longmapsto w \otimes v. \end{aligned}$$

An important example of a monoidal category that admits more than one symmetry is

3.2.46 The category of graded vector spaces. A *graded vector space* is a direct sum of vector spaces $V = \bigoplus_{n\in\mathbb{Z}} V_n$, and a graded linear map is one that respects the grading. The tensor product of two graded vector spaces V and W is again a graded vector space, with grading $(V \otimes W)_n = \bigoplus_{p+q=n}(V_p \otimes W_q)$. The ground field $\Bbbk$ is a graded vector space concentrated in degree 0. So we have a monoidal category $(\boldsymbol{grVect}_{\Bbbk}, \otimes, \Bbbk)$ of graded vector spaces. Now of course we have the canonical symmetry σ just as above: $v \otimes w \mapsto w \otimes v$. But there is also another important possibility for defining a twist map, namely via Koszul's sign change:

$$v \otimes w \mapsto (-1)^{pq}\, w \otimes v$$

where $\deg(v) = p$ and $\deg(w) = q$. One checks that this symmetry κ does indeed satisfy the axioms, making $(\boldsymbol{grVect}_{\Bbbk}, \otimes, \Bbbk, \kappa)$ into a symmetric monoidal category.

3.2.47 Strictification of symmetry? Why not forget about the symmetry 2-cell and plainly regard it as an identity, just as we do for the 2-cells in the definition of nonstrict monoidal categories? This is tempting, and sometimes you can get away with it (for example Quinn takes this viewpoint in [43], and writes $V \otimes W = W \otimes V$, for any two vector spaces V, W). However this is a delicate question, and in general this simplification will not work: there is no strictification result like 'every symmetric monoidal category is equivalent to a strictly commutative one'. (But there is a coherence result, cf. Mac Lane [34], new edition, Chapter XI.)

A more prosaic reason for not pretending that symmetries are identities is the above example, the category of graded vector spaces: we would confuse ourselves completely if we regarded the symmetry κ as the identity!

3.2.48 Symmetric monoidal functors. Given two symmetric monoidal categories $(\mathbf{V}, \square, I, \tau)$ and $(\mathbf{V}', \square', I', \tau')$, an obvious property to require from a monoidal functor $F : \mathbf{V} \to \mathbf{V}'$ is that it preserve the symmetric structure, namely that for every pair of objects X, Y in $\mathbf{V}$ we have

$$\tau_{X,Y} F = \tau'_{XF,YF}.$$

In other words, the image of the twist map is the twist map of the images. Such monoidal functors are called *symmetric monoidal functors*. The composition of two symmetric monoidal functors is clearly symmetric again, and the identity monoidal functor is symmetric. All told, there is a category ***SymMonCat*** of symmetric monoidal categories and symmetric monoidal functors.

Monoidal functor categories

3.2.49 Monoidal natural transformations. Let $(\boldsymbol{V}, \Box, I)$ and $(\boldsymbol{V}', \Box', I')$ be two monoidal categories, and let

$$\boldsymbol{V} \underset{F}{\overset{G}{\rightrightarrows}} \boldsymbol{V}'$$

be two monoidal functors. A natural transformation $u : F \Rightarrow G$ is called a *monoidal natural transformation* if for every two objects X, Y in $\boldsymbol{V}$ we have

$$u_X \Box' u_Y = u_{X \Box Y},$$

and also $u_I = \mathrm{id}_{I'}$. The first requirement makes sense because $(X \Box Y)F = XF \Box' YF$ and $(X \Box Y)G = XG \Box' YG$, so we can write

$$\begin{array}{ccc} (X\Box Y)F & \xrightarrow{u_{X\Box Y}} & (X\Box Y)G \\ \| & & \| \\ XF\Box' YF & \xrightarrow[u_X \Box' u_Y]{} & XG \Box' YG \end{array}$$

The second requirement makes sense because $IF = IG = I'$.

3.2.50 Monoidal functor categories. For two fixed monoidal categories $(\boldsymbol{V}, \Box, I)$ and $(\boldsymbol{V}', \Box', I')$, there is a category ***MonCat***$(\boldsymbol{V}, \boldsymbol{V}')$ whose objects are the monoidal functors from $\boldsymbol{V}$ to $\boldsymbol{V}'$, and whose arrows are the monoidal natural transformations between such functors.

3.2.51 Symmetric monoidal functor categories. Similarly, given two symmetric monoidal categories $(\boldsymbol{V}, \Box, I, \tau)$ and $(\boldsymbol{V}', \Box', I', \tau')$, there is a category ***SymMonCat***$(\boldsymbol{V}, \boldsymbol{V}')$ whose objects are the symmetric monoidal functors from $\boldsymbol{V}$ to $\boldsymbol{V}'$, and whose arrows are the monoidal natural transformations between such functors.

For our purposes, a particularly important class of monoidal functors are the

3.2.52 Linear representations. By definition, a *linear representation* of a symmetric monoidal category $(\boldsymbol{V}, \Box, I, \tau)$ is a symmetric monoidal functor $(\boldsymbol{V}, \Box, I, \tau) \to (\mathbf{Vect}_{\Bbbk}, \otimes, \Bbbk, \sigma)$, where σ is the usual symmetry (cf. 3.2.45). So the set of all linear representations of $\boldsymbol{V}$ is the objects of a category which we denote

$$\boldsymbol{Repr}_{\Bbbk}(\boldsymbol{V}) := \boldsymbol{SymMonCat}(\boldsymbol{V}, \mathbf{Vect}_{\Bbbk}).$$

3.2.53 Note. We consider *strict* functors. In fact our ***Vect*** is a strictification of the true monoidal category of vector spaces… Otherwise we should consider *strong* monoidal functors, cf. 3.2.17…

3.2.54 TQFTs. By definition 1.3.32, a topological quantum field theory is a symmetric monoidal functor from $(\boldsymbol{nCob}, \coprod, \emptyset, T)$ to $(\boldsymbol{Vect}_{\Bbbk}, \otimes, \Bbbk, \sigma)$. Such functors form the objects of a category

$$\boldsymbol{nTQFT}_{\Bbbk} = \boldsymbol{Repr}_{\Bbbk}(\boldsymbol{nCob}) = \boldsymbol{SymMonCat}(\boldsymbol{nCob}, \boldsymbol{Vect}_{\Bbbk}),$$

the arrows being the monoidal natural transformations.

Exercises

1. Let $\boldsymbol{Mat}_{\Bbbk}$ denote the category whose object set is $\mathbb{N}$, and whose arrows are matrices over $\Bbbk$, an m-by-n matrix being regarded as an arrow from m to n. Composition of arrows is matrix multiplication. Show that $\boldsymbol{Mat}_{\Bbbk}$ becomes a monoidal category under the operation of addition in $\mathbb{N}$, and forming block matrices like this: $A \Box B := \left(\begin{smallmatrix} A & 0 \\ 0 & B \end{smallmatrix}\right)$. (For this to work you have to invent 0-by-n matrices and m-by-0 matrices…)
2. Show that if ϕ and ϕ' are invertible arrows in a monoidal category $(\boldsymbol{V}, \Box, I)$, then $\phi \Box \phi'$ is also invertible.
3. Let $(\boldsymbol{V}, \Box, I)$ be a monoidal category. The goal of the exercise is to show that the endomorphism monoid $\mathrm{End}_{\boldsymbol{V}}(I)$ is commutative.
   Independently of the monoidal structure, the set $\mathrm{End}_{\boldsymbol{V}}(I)$ is naturally a monoid under composition of arrows (cf. 3.1.18); the unit is $\mathrm{id}_I : I \to I$. Show that there is a second monoid structure on $\mathrm{End}_{\boldsymbol{V}}(I)$ obtained from $\Box$ via the identification $I \Box I = I$. Namely, given $f : I \to I$ and $g : I \to I$ we can consider $f \Box g : I \Box I \to I \Box I$ as an arrow $I \to I$. Show that id_I is the unit for this monoid structure. Next, decompose $f \Box g$ with the help of the identity arrow as in 3.2.6 and use these decompositions to show that $f \Box g$ can be identified with fg as well as with gf. (This is one form of the Eckmann–Hilton argument. A more general version is given in Exercise 2 on page 211.)
4. Let $\boldsymbol{C}$ be any category and consider the category $\mathrm{End}_{\boldsymbol{Cat}}(\boldsymbol{C})$ (also denoted $\boldsymbol{Cat}(\boldsymbol{C}, \boldsymbol{C})$) whose objects are functors $\boldsymbol{C} \to \boldsymbol{C}$ and whose arrows are natural transformations. Show that composition of functors (and the identity functor) make $\mathrm{End}_{\boldsymbol{Cat}}(\boldsymbol{C})$ into a monoidal category. (This is a categorification of 3.1.18.)
5. Let ***Boole*** denote 'the category of Boolean logic' in this sense: the objects are propositions (say $p, q, \dots$), and there is an arrow from p to q whenever

p implies q. (So between two objects there is either one or zero arrows.) Show that $\wedge$ = AND is a categorical product and that $\top$ = TRUE is a terminal object. (Hence (by 3.2.22 and 3.2.42), (***Boole***, $\wedge$, $\top$) is a symmetric monoidal category.) Analogously, show that $\vee$ = OR is a categorical coproduct and that $\perp$ = FALSE is an initial object. (Hence (***Boole***, $\vee$, $\perp$) is a symmetric monoidal category.)

Let **2** $\subset$ ***Boole*** denote the full subcategory consisting of $\top$ and $\perp$ (so there is a single nonidentity arrow $\perp \to \top$). Show that the two monoidal structures on ***Boole*** induce monoidal structures on **2**. (See also A.1.4 and Exercise 1 on page 233.)

6. Consider the two monoidal categories (***Set***, $\coprod$, $\emptyset$) and ($\boldsymbol{Vect}_{\Bbbk}$, $\oplus$, 0). Show that the free functor ***Set*** $\to$ $\boldsymbol{Vect}_{\Bbbk}$ studied in Exercise 1 on page 92 is a monoidal functor with respect to these two monoidal structures.
   Consider now instead the two monoidal categories (***Set***, $\times$, 1) and ($\boldsymbol{Vect}_{\Bbbk}$, $\otimes$, $\Bbbk$). Show that the free functor is monoidal with respect to these two monoidal structures.
7. Consider now the monoidal categories (***Set***, $\coprod$, $\emptyset$) and ($\boldsymbol{Alg}_{\Bbbk}$, $\otimes$, $\Bbbk$). Show that the free-algebra functor ***Set*** $\to$ $\boldsymbol{Alg}_{\Bbbk}$ defined in Exercise 7 on page 93 is a monoidal functor with respect to these two monoidal structurcs.
8. (Joyal and Street [28].) A *braiding* on a monoidal category ($\boldsymbol{V}$, $\Box$, I) is a family of maps $\tau_{X,Y} : X \Box Y \to Y \Box X$ subject to axioms (i) and (ii) on page 161 – but not axiom (iii). A *braided monoidal category* is a monoidal category equipped with a braiding. We picture (the components of) a braiding like this

   Copy over the arguments in 3.2.36–3.2.39 to show that this equation holds for any three objects in a braided monoidal category:

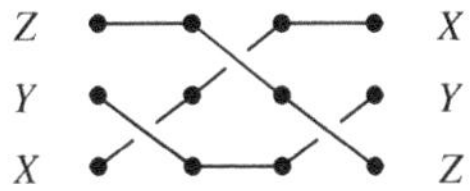

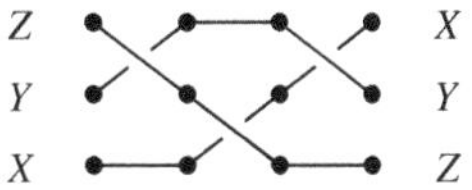

   Prove also the analogue of Lemma 3.2.40.
9. Let G be a group. A *G-set* is a set X together with a right G-action (cf. 3.1.21)

$$X \times G \longrightarrow X$$
$$(x, g) \longmapsto x.g.$$

A *homomorphism of G-sets* (also called a *G-map*) is a set map $\phi : X \to Y$ compatible with the actions, i.e. $(x.g)\phi = (x\phi).g$. Let $\mathbf{Set}_G$ denote the category of G-sets and G-maps. Show that this category is monoidal under the operation that sends a pair of G-sets X and Y to the product set $X \times Y$ with coordinate-wise G-action

$$\begin{aligned} X \times Y \times G &\longrightarrow X \times Y \\ (x, y, g) &\longmapsto (x.g, y.g). \end{aligned}$$

10. (Freyd and Yetter.) Continuing the previous exercise, a particular G-set is the set underlying G itself, with action given by conjugation:

$$\begin{aligned} G \times G &\longrightarrow G \\ (x, g) &\longmapsto g^{-1}xg. \end{aligned}$$

A *crossed G-set* is a G-set X equipped with a G-map $X \to G$, which we denote $x \mapsto |x|$. So it means that this square commutes:

$$\begin{array}{ccc} X \times G & \longrightarrow & X \\ \downarrow & & \downarrow \\ G \times G & \longrightarrow & G \end{array}$$

which in turn is expressed in terms of elements

$$|x.g| = g^{-1}\,|x|\,g.$$

A *crossed G-map* $\phi : X \to Y$ between two crossed G-sets is a G-map compatible with the maps to G. That is, $|x\phi| = |x|$, for all $x \in X$. Let us denote by $\mathbf{Set}_G/G$ the category of crossed G-sets and crossed G-maps. Show that the map

$$\begin{aligned} X \times Y &\longrightarrow G \\ (x, y) &\longmapsto |x|\,|y| \end{aligned}$$

defines monoidal structure on $\mathbf{Set}_G/G$.

11. For each pair X, Y of crossed G-sets, put

$$\begin{aligned} \tau_{X,Y} : X \times Y &\longrightarrow Y \times X \\ (x, y) &\longmapsto (y, x.|y|), \end{aligned}$$

and show that this defines a braiding on $\mathbf{Set}_G/G$.

3.3 Frobenius algebras and 2-dimensional topological quantum field theories

3.3.1 2-Dimensional TQFTs and Frobenius algebras. So a 2-dimensional TQFT is a linear representation of the symmetric monoidal category $(\textbf{2Cob}, \coprod, \emptyset, T)$. This is a category we fully control, because we are given a presentation of it in terms of generators and relations. Recall that the objects of ***2Cob*** are $\{\mathbf{0}, \mathbf{1}, \mathbf{2}, \dots\}$ where $\mathbf{n}$ is the disjoint union of n circles, and that the generating arrows are

(The precise meaning of these symbols was given in 1.4.) Recall furthermore that we found a bunch of relations, the most important ones being

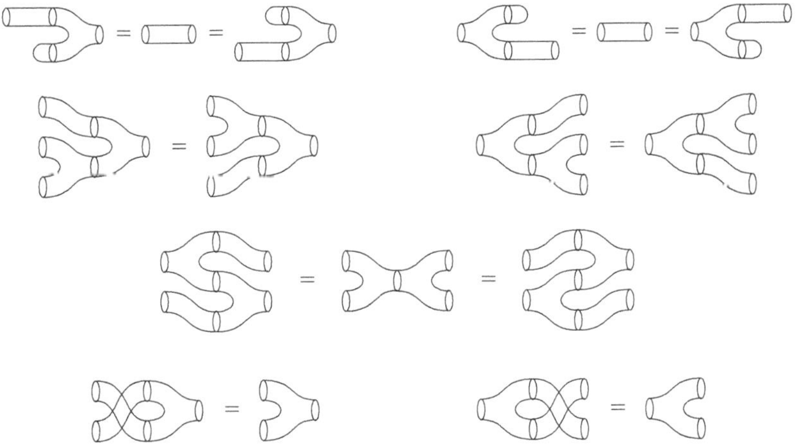

(There was furthermore a jungle of relations involving the twist map (1.4.35) which just amount to saying that ***2Cob*** is a symmetric monoidal category. Since we require our monoidal functor to be symmetric, all these relations are automatically taken care of, so we do not have to bother with them.)

In general, a monoidal functor is determined completely by its values on the generators of the source category. In our case we want to specify a symmetric monoidal functor $\mathscr{A} : \textbf{2Cob} \to \textbf{Vect}_{\mathbb{k}}$, so we must specify a vector space A as image of $\mathbf{1}$, and a linear map for each of the generators. The fact that the functor is monoidal implies in particular that the image of $\mathbf{2}$ is $A \otimes A$, and so on. To ease the notation, put $A^n := A \otimes \dots \otimes A$ (with n factors). The fact that $\mathscr{A}$ is a *symmetric* monoidal functor means that the image of [twist] must be the usual twist for the tensor product, so the following is automatic once we have fixed the vector space$A$:

$$\begin{aligned}
\textbf{\textit{2Cob}} &\longrightarrow \textbf{\textit{Vect}}_{\Bbbk} \\
\mathbf{1} &\longmapsto A \\
\mathbf{n} &\longmapsto A^n \\
&\longmapsto [\mathrm{id}_A : A \to A] \\
&\longmapsto [\sigma : A^2 \to A^2].
\end{aligned}$$

Let the images of the generators be denoted like this:

$$\begin{aligned}
\textbf{\textit{2Cob}} &\longrightarrow \textbf{\textit{Vect}}_{\Bbbk} \\
&\longmapsto [\eta : \Bbbk \to A] \\
&\longmapsto [\mu : A^2 \to A] \\
&\longmapsto [\varepsilon : A \to \Bbbk] \\
&\longmapsto [\delta : A \to A^2].
\end{aligned}$$

So A is a vector space equipped with certain linear maps among its tensor powers; now the *relations* that hold in ***2Cob*** translate into relations among these linear maps. With the graphical notation we used in Chapter 2 we anticipated this comparison: it is easy to see that the relations translate exactly into the axioms for a commutative Frobenius algebra.

So in conclusion, given a 2-dimensional TQFT $\mathscr{A}$, then the image vector space $A = \mathbf{1}\mathscr{A}$ is a commutative Frobenius algebra.

Conversely, starting with a commutative Frobenius algebra (A, ε) (whose multiplication is denoted μ, etc.) then we can construct a TQFT $\mathscr{A}$ by using the above description as definition. In other words, $\mathscr{A}$ is defined by sending $\mathbf{1} \mapsto A$ (the underlying vector space), $\mapsto \mu$, etc. Here we must check that this makes sense of course – that the relations are respected. For example we must check that for the cobordism

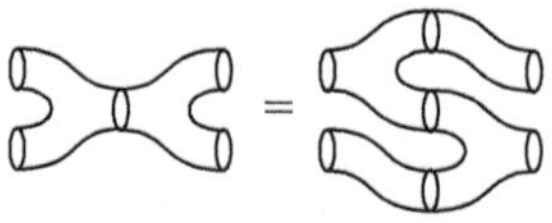

$$=$$

it makes no difference whether we let the image be $\mu\delta$ or $(\delta \otimes \mathrm{id}_A)(\mathrm{id}_A \otimes \mu)$. But again, since the relations in ***2Cob*** correspond precisely to the axioms for a commutative Frobenius algebra, this is automatic, so the symmetric monoidal functor $\mathscr{A}$ is indeed well defined.

Also it is clear that these two constructions are inverse to each other: e.g., if we start with $\mathscr{A}$ and construct a commutative Frobenius algebra $A := \mathbf{1}\mathscr{A}$, and then define a symmetric monoidal functor such that $\mathbf{1} \mapsto A$, then we recover $\mathscr{A}$. So we have established a one-to-one correspondence between 2-dimensional TQFTs and commutative Frobenius algebras.

(Beware: we have certain *symmetric* monoidal functors on one side, and *commutative* Frobenius algebras on the other side, but the adjectives 'symmetric' and 'commutative' do not correspond precisely to each other here. See 3.3.3 for further discussion.)

This correspondence also works for arrows. The arrows in $\textbf{\textit{2TQFT}}_{\Bbbk}$ are the monoidal natural transformations. Given two TQFTs $\mathscr{A}$, $\mathscr{B}$, i.e. two symmetric monoidal functors $\textbf{\textit{2Cob}} \to \textbf{\textit{Vect}}_{\Bbbk}$, then a natural transformation u between them consists of linear maps $A^n \to B^n$ for each $n \in \mathbb{N}$. That u is a *monoidal* natural transformation means that the map $A^n \to B^n$ is just the nth tensor power of the map $A^1 \to B^1$, so u is determined completely by specifying this linear map. The *naturality* of u means that all these maps are compatible with arrows in $\textbf{\textit{2Cob}}$. Now every arrow in $\textbf{\textit{2Cob}}$ is built up from the generators, so naturality boils down to four commutative diagrams (one for each generator). For example the diagrams corresponding to and are

which amounts precisely to the statement that $A \to B$ is a $\Bbbk$-algebra homomorphism. Similarly, the conditions corresponding to and express that $A \to B$ is a coalgebra homomorphism. So all together it is then a Frobenius algebra homomorphism (cf. the definition 2.4.4). (The 'generator' is in fact not a generator at all, and the diagram one could draw involving it commutes automatically. Finally, is a true generator, but since the two monoidal functors are required to be symmetric they preserve twist maps, so there is no interesting diagram to draw for this one either.)

Conversely, given a Frobenius algebra homomorphism between two commutative Frobenius algebras, we can use the above arguments in the reverse direction to construct a monoidal natural transformation between the TQFTs corresponding to A and B.

All together we have proved our main theorem:

3.3.2 Theorem. *There is a canonical equivalence of categories*

$$\textbf{\textit{2TQFT}}_{\Bbbk} \simeq \textbf{cFA}_{\Bbbk}.$$

If we regard $\mathbf{2TQFT}_{\Bbbk}$ as the category of representations of the skeletal version of $\mathbf{2Cob}$, then in fact we have an *isomorphism* of categories. (See Remark 3.5.19 for further discussion of such issues.)

3.3.3 What about the symmetry requirement? It is natural to ask whether our requirement that the monoidal functor defining a TQFT be symmetric is desirable or necessary – it might turn out to be symmetric automatically?!

Here is an example which illuminates this question, and in particular shows that the symmetry of the functor is not automatic. For a moment, let us drop the requirement that a TQFT be symmetric. Consider a *graded-commutative* Frobenius algebra H, for example the cohomology ring of a compact manifold (cf. 2.2.23). Now define a nonsymmetric TQFT by sending $\mathbf{1} \mapsto H$: Concerning the generators for $\mathbf{2Cob}$, send them to multiplication and unit, comultiplication and counit, just as usual, but send the twist cobordism to Koszul's sign-changed twist (cf. 3.2.46)

$$a \otimes b \mapsto (-1)^{pq}\, b \otimes a,$$

where $\deg(a) = p$ and $\deg(b) = q$. This works! Even though H is not a *commutative* Frobenius algebra in the usual sense, there is no contradiction because the twist cobordism is not sent to the usual twist!

So here we have a natural example of a good monoidal functor $\mathbf{2Cob} \to \mathbf{Vect}_{\Bbbk}$ which is not symmetric, but unfortunately our current definition of TQFT (with the symmetry requirement) is too narrow to include it. What should we do with this example? If we drop the symmetry requirement, Theorem 3.3.2 would simply be wrong! – we get noncommutative Frobenius algebras as well. Then you might try to formulate the theorem without symmetry requirement using more generally graded-commutative Frobenius algebras. But how can you be sure there is not yet another 'exotic-commutative algebra' notion defying your theorem, just as the graded-commutative example defies the version of the theorem without symmetry requirement? To be sure, the formulation of the theorem would be something like: '(not necessarily symmetric) TQFTs correspond to such Frobenius algebras which are exotic-commutative in some sense (relating to some symmetric structure on some **Vect**-like monoidal category in which they can be considered to live)'.

In any case, it is clear that symmetry exists on both sides of the functor $\mathscr{A}$, and morally of course it is wrong to deny it or ignore it. The ugliness of the last statement is convincing evidence that we are doing something wrong.

The point is that commutativity is actually a relative notion: it depends on a symmetry structure on the ambient monoidal category (cf. 3.5.6). If instead of the usual $\mathbf{Vect}_{\Bbbk}$ we place ourselves in the symmetric monoidal category

$(\boldsymbol{grVect}_{\Bbbk}, \otimes, \Bbbk, \kappa)$, then the algebra H *is* commutative, and the monoidal functor $\mathbf{1} \mapsto H$ *is* symmetric. So the good way for the theorem to handle this tricky example is not to give up symmetry – on the contrary: *symmetric monoidal* is the crucial notion here, so we need to consider more general symmetric monoidal categories – $(\boldsymbol{Vect}_{\Bbbk}, \otimes, \Bbbk, \sigma)$ should not have a monopoly of receiving the functors defining TQFTs! Also, from the viewpoint of quantum theory, there is no reason for favouring precisely the category $(\boldsymbol{Vect}_{\Bbbk}, \otimes, \Bbbk, \sigma)$ – in fact, in the list of examples that Atiyah gives in [5], half of the examples really use mod-2 graded vector spaces instead of plainly vector spaces – such creatures abound in quantum physics...

So this whole discussion is one motivation for the step of abstraction we will take in the remainder of these notes, allowing arbitrary (symmetric monoidal) target categories. Another motivation is that in fact it is no more difficult to treat this general case.

3.3.4 Historical remarks. The observation that 2-dimensional TQFTs are essentially the same thing as commutative Frobenius algebras was first made by R. Dijkgraaf in his Ph.D. thesis [16]. More precise proofs have been given by Dubrovin [19], Quinn [43], Sawin [44], and Abrams [1] – this is at the same time the chronological order and the order according to the amount of detail presented. However, all these sources are silent on the questions of symmetry...

3.3.5 What's next? With Theorem 3.3.2 we have finished: we understand the relation between 2-dimensional TQFTs and Frobenius algebras! and in the exercises we work out a couple of simple examples. The rest of these notes is devoted to placing the above theorem in its proper context. We will show that it is just a variation of a more basic result: there is a monoidal category Δ (the simplex category) which is quite similar to ***2Cob*** (in fact it is a subcategory) such that giving a monoidal functor from Δ to $\boldsymbol{Vect}_{\Bbbk}$ is the same as giving an algebra:

$$\boldsymbol{MonCat}(\Delta, \boldsymbol{Vect}_{\Bbbk}) \simeq \boldsymbol{Alg}_{\Bbbk}.$$

This in turn is just a special case of a general principle (another example being that monoidal functors $\Delta \to \boldsymbol{Ab}$ are just rings). The general result states that monoidal functors from Δ to any monoidal category $\boldsymbol{V}$ correspond to monoids in $\boldsymbol{V}$. This amounts to saying that Δ is the free monoidal category on a monoid.

This result also has a variant for Frobenius algebras: we will define a notion of *Frobenius object* in a general monoidal category (such that a Frobenius object in $\mathit{\mathbf{Vect}}_{\Bbbk}$ is precisely a Frobenius algebra). Similarly, a commutative Frobenius object in a symmetric monoidal category is the notion such that a commutative Frobenius object in $(\mathit{\mathbf{Vect}}_{\Bbbk}, \otimes, \Bbbk, \sigma)$ is precisely a commutative Frobenius algebra. We will see that ***2Cob*** does for commutative Frobenius objects what Δ does for monoids: every commutative Frobenius object (in any symmetric monoidal category) arises as the image of a unique symmetric monoidal functor from ***2Cob***. In other words, ***2Cob*** *is the free symmetric monoidal category on a commutative Frobenius object* (Theorem 3.6.19).

Our main motivation for striving for this generality is to place Theorem 3.3.2 in its natural context, isolating those properties of $\mathit{\mathbf{Vect}}_{\Bbbk}$ that we used. Example 3.3.3 hints at the importance of this generality: although the category of vector spaces (with its canonical symmetry) is important, it is not the only interesting symmetric monoidal category, and it is worth looking at TQFTs with values in other monoidal categories.

Exercises

1. (Sawin [44].) Show that for 2-dimensional TQFTs, the direct sum notion of Durhuus and Jónsson (Exercise 4 on page 56) corresponds exactly to the notion of direct product of Frobenius algebras (cf. Exercise 7 on page 105).

2-Dimensional TQFTs (i.e. commutative Frobenius algebras) fall in two major groups: nilpotent and semi-simple – and then by taking direct sum you can mix the two types. These behave quite differently. In the next couple of exercises we just work out what sorts of invariants of surfaces they produce. Recall from 1.2.29 that each closed surface is considered as a cobordism $\varnothing \Rightarrow \varnothing$, so its image under a TQFT is a linear map $\Bbbk \to \Bbbk$, i.e. a constant, which is a topological invariant of the surface.

2. (Typical nilpotent example.) Let $\mathscr{A}$ be the 2-dimensional TQFT corresponding to the Frobenius algebra $A = \Bbbk[t]/t^n$, with Frobenius form $t^{n-1} \mapsto 1$, other generators t^i mapping to zero (see 2.2.21). Show that the invariants of the closed connected surfaces are as follows: the sphere has invariant 0, the torus has invariant n, and all surfaces of higher genus have invariant 0. (Hint: use Exercise 11 on page 129.) So in conclusion this TQFT does not produce very fine invariants!

   Compute also the invariants of all nonconnected surfaces.

3. (Typical semi-simple example.) Let $\mathscr{A}$ be the 2-dimensional TQFT corresponding to the group algebra of the cyclic group of order n (cf. 2.2.18). That is, $A = \Bbbk[t]/(t^n - 1)$, and as Frobenius form we take $1 \mapsto 1$, other generators t^i mapping to zero. Show that this TQFT associates invariant n^g

to the closed connected surface of genus g. So this TQFT can detect genus (assuming the characteristic of $\mathbb{k}$ does not divide n).
What is the invariant of the disjoint union of two genus-g surfaces?

4. As a variation of the preceding example, take $n = 2$: then our algebra is $A = \mathbb{k}[t]/(t^2 - 1)$, but let now the Frobenius form be $t \mapsto 1$, $1 \mapsto 0$. Show that the handle element is $2t$, and conclude that the corresponding TQFT associates invariant 0 to any closed connected surface of even genus!
What about the disjoint union of two tori?
5. Compute the invariants produced by the TQFT obtained from the direct sum of $\mathbb{k}[t]/t^2$ and $\mathbb{k}[t]/(t^2 - 1)$, with Frobenius forms as in Exercise 2 and Exercise 3.
6. Consider the trivial Frobenius algebra $\mathbb{k}$ (with Frobenius form $\mathbb{k} \to \mathbb{k}$, $1 \mapsto 1$), cf. 2.2.12. Show that the corresponding TQFT gives invariant 1 to every surface.
7. Consider now the 1-dimensional Frobenius algebra $\mathbb{k}$ over $\mathbb{k}$, with Frobenius form $1 \mapsto 2$. Show that the corresponding TQFT sends the disjoint union of k spheres to the invariant 2^k, and does not detect any surfaces of positive genus. (So it is a sphere-counting TQFT.)
8. Use the cyclic group of order 31 and a field of characteristic 31 to construct a TQFT that can distinguish between any number of spheres up to 30, but gives zero if any surface of higher genus appears.

3.4 The simplex categories Δ and Φ

Finite ordinals

3.4.1 The category of finite (totally) ordered sets. An *ordering* on a finite set S is a relation $\leq$ which is transitive, reflexive, and anti-symmetric. Transitivity means that $a \leq b \leq c$ implies $a \leq c$; reflexivity means $a \leq a$ for all $a \in S$; and anti-symmetry means that $a \leq b \leq a$ implies $a = b$. The ordering is *total* if furthermore for each a, b we have $a \leq b$ or $b \leq a$. The empty set is a special example of a finite set, and it is automatically totally ordered since there are no elements in it!

All our orderings will be total so we will suppress the adjective 'total': from now on, 'ordered' means 'totally ordered'.

An *order-preserving map* between finite ordered sets is a map $f : S \to S'$ such that $a \leq b$ in S implies $af \leq bf$ in S'. Clearly the composition of two order-preserving maps is again order preserving, and clearly the identity map on any finite ordered set is order preserving, so there is a category ***FinOrd*** of finite (totally) ordered sets and order-preserving maps.

3.4.2 Alternative categorical description. Consider a category $\boldsymbol{S}$ with finitely many objects, and such that given two objects there is precisely one arrow between them – *either* in one direction *or* in the other direction. Such a category is essentially the same as a finite (totally) ordered set: starting from the finite ordered set S construct a category $\boldsymbol{S}$ by taking the elements of S as objects, and introducing an arrow from a to b if and only if $a \leq b$ in S. Then transitivity defines composition in the category $\boldsymbol{S}$; reflexivity means that we have an identity arrow for each object, and anti-symmetry says that there cannot be arrows in both directions between two distinct objects in $\boldsymbol{S}$. Conversely, the same arguments show that such a category defines a finite ordered set. Concerning maps: the order-preserving maps between finite ordered sets correspond exactly to the functors between the categories.

We will be more explicit about this viewpoint in 3.4.5.

3.4.3 Monoidal structure on *FinOrd*. The category of finite ordered sets becomes a monoidal category under disjoint union, which we will denote as $+$. Clearly the disjoint union of two finite ordered sets S, S' is again a finite set, but we have to specify an order on the resulting finite set $S + S'$. We do that by declaring that the elements of S 'come before' the elements of S'. Precisely: for any $x \in S$ and $x' \in S'$ we declare $x \leq x'$. The neutral element for $+$ is clearly the empty set $\varnothing$. We should note that this monoidal category is not symmetric (i.e. there does not exist a symmetric structure on it). We will see that in 3.4.9.

There are two slightly annoying things with the category of finite ordered sets. First, this is a very big category. Second, the notion of disjoint union is only well defined up to canonical isomorphism. We can get a more manageable category by taking a skeleton. Recall that a skeleton of a category is a full subcategory comprising exactly one object from each isomorphism class.

3.4.4 The category Δ of finite ordinals (also called the simplex category). Since two finite ordered sets are isomorphic if and only if they have the same number of elements, to construct a skeleton for ***FinOrd*** we need to choose one ordered set for each $n \in \mathbb{N}$. A representative is called a *finite ordinal*. To be specific, for each $n \in \mathbb{N}$, let $\mathbf{n}$ denote the ordered set $\{0, \ldots, n-1\}$. So in particular, $\mathbf{0}$ is the empty set and $\mathbf{1}$ is the one-element set $\{0\}$. Now our skeleton of ***FinOrd*** will be given by

$$\Delta = \{\mathbf{0}, \mathbf{1}, \mathbf{2}, \ldots\}.$$

The arrows of Δ are the order-preserving maps between those sets. In other words, they are functions $f : \mathbf{m} \to \mathbf{n}$ such that $i \leq j$ in $\mathbf{m}$ implies $if \leq jf$ in $\mathbf{n}$.

Now we argue that the ordinal sum makes Δ into a (strict) monoidal category. The *ordinal sum* $\mathbf{m} + \mathbf{n}$ of two ordinals $\mathbf{m}$ and $\mathbf{n}$ is simply the ordinal corresponding to the natural numbers sum. The set $\mathbf{m} + \mathbf{n}$ already comes with an order, but we must specify how the two original sets inject into it: these (order-preserving) injections are given by

$$\begin{aligned} \mathbf{m} &\longrightarrow \mathbf{m} + \mathbf{n} \\ i &\longmapsto i \end{aligned}$$

$$\begin{aligned} \mathbf{n} &\longrightarrow \mathbf{m} + \mathbf{n} \\ j &\longmapsto m + j. \end{aligned}$$

We also have to specify what the composition $+$ does to arrows. If $f : \mathbf{m} \to \mathbf{n}$ and $f' : \mathbf{m}' \to \mathbf{n}'$ are two order-preserving maps, then the function $f + f'$ from $\mathbf{m} + \mathbf{m}'$ to $\mathbf{n} + \mathbf{n}'$ is defined as

$$\begin{aligned} \mathbf{m} + \mathbf{m}' &\longrightarrow \mathbf{n} + \mathbf{n}' \\ i &\longmapsto \begin{cases} if & \text{for } i = 0, \ldots, m-1 \\ n + (i-m)f' & \text{for } i = m, \ldots, m + m' - 1. \end{cases} \end{aligned}$$

3.4.5 Δ as subcategory of *Cat*. We can give a categorical version of the definition, by interpreting the order as arrows in a category: let $\mathbf{n}$ denote the category whose objects are $\{0, 1, \ldots, n-1\}$ and whose arrows are the order relation, so there is exactly one arrow $i \to j$ whenever $i \leq j$. In particular $\mathbf{0}$ is the category which has no objects and no arrows! $\mathbf{1}$ is the category with only one object and its identity arrow, in accordance with the notation used elsewhere in these notes. $\mathbf{2}$ is the category which we could picture as $\bullet \to \bullet$: there are two objects and one arrow (in addition to the identity arrows which we have not drawn). Next, $\mathbf{3}$ is the category generated by $\bullet \to \bullet \to \bullet$. In addition to these two arrows there is a third arrow which we have not drawn, namely their composition, and of course the three identity arrows. Next, $\mathbf{4}$ is generated by a chain of three arrows; with the possible compositions you arrive at the graph of a tetrahedron. In general, $\mathbf{n}$ is the oriented graph of an n-simplex!

Now we can characterise Δ as the full subcategory of ***Cat*** whose objects are $\mathbf{0}, \mathbf{1}, \mathbf{2}, \ldots$ Indeed, since the arrows in these categories are just the orderings, a functor between two such categories is just what before we called an order-preserving map.

Graphical description of Δ

3.4.6 Objects and arrows. Define a category as follows. The *objects* are finite sets of dots arranged in a column like this

We also allow the empty column, so there is an object for each $n \in \mathbb{N}$.

An *arrow* from one column to another is a collection of strands starting at the dots in the source column and ending at dots in the target column, subject to the following two rules:

(i) for each dot in the source column there is exactly one strand coming out (and going to the target column);

(ii) the strands are not allowed to cross each other, but they are allowed to merge – in other words, two or more strands may share a single dot in the target.

Here is an example:

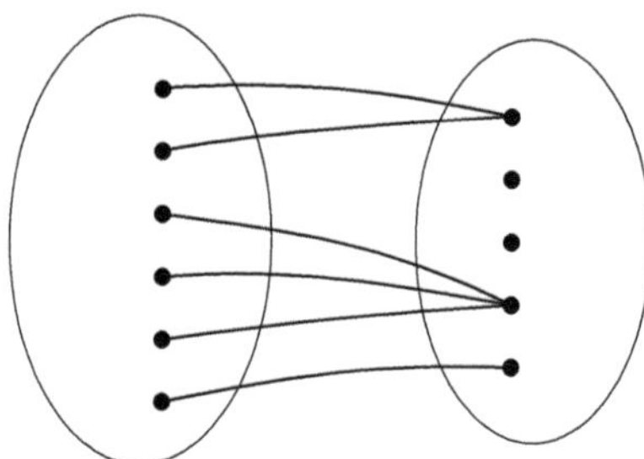

Actually some fine print is needed to avoid misinterpretation of these rules. First of all, we should require the strands to leave the input column rightwards and arrive at the output column from the left. Otherwise we could imagine figures like these

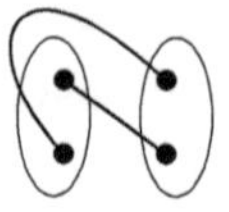
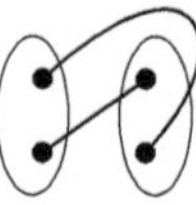

which would defy the intention of rule (ii). Also, in order for rule (ii) to have any effect at all, we should require the entire drawing to be in the plane – not in 3-space or on a torus or anything wild like that. To be concrete, we could require each arrow to take place inside a rectangle; then the input column should sit on the left-hand side of the rectangle and the output column on the right-hand side – like this:

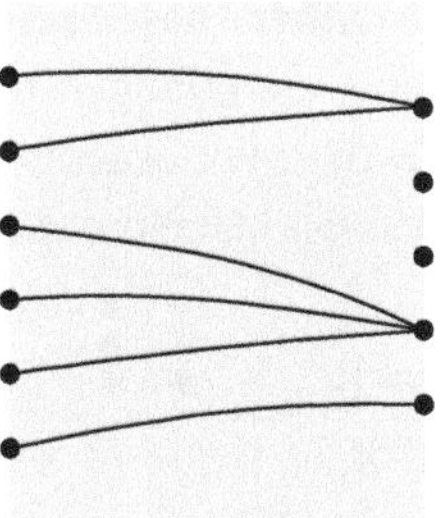

Finally, we should specify when we consider two collections of strands to define the same arrow – the criterion is this: it only matters *where* a dot goes, not *how* it gets there. So the actual shape of the strand is immaterial: we could say that two strands are equivalent if one can be continuously deformed into the other (inside the rectangle and without touching the other strands), and then say that equivalent strands define the same arrow. (In fancier terminology we are talking about their *isotopy classes*.)

3.4.7 Composition and neutral arrows. The *composition* of two arrows is given by joining the input ends of the second collection of strands to the output ends of the first collection of strands – provided the columns match, of course. Like this:

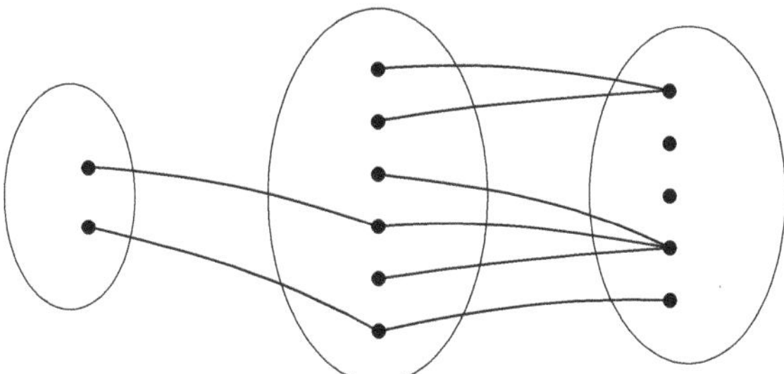

(Observe that composition does not lead to any crossing-over, thanks to the fine print condition. Also, since each of the arrows takes place in a rectangle, the composition can be viewed as taking place in the rectangle obtained by gluing together the two original rectangles along the output side of the first and the input side of the second.)

For each object there is an identity arrow given by taking a strand from each dot to itself. It is easy to check that these definitions satisfy the axioms for a category. In fact we can see this quite easily in the following way. The first rule amounts to saying that f is a *function* from the set of dots in the source column to the set of dots in the target column. This picture is in all books on sets and functions. Then composition of arrows is just composition of functions, and the identity arrows are simply the identity functions.

3.4.8 Comparison with the other descriptions of Δ, and monoidal structure. The fact that the dots are arranged in columns is just to say that the sets are ordered, and rule (ii) says exactly that the functions are order preserving. So we are really just talking about Δ; our new description is

$$\Delta = \{\varnothing,\ \bullet,\ \substack{\bullet\\ \bullet},\ \substack{\bullet\\ \bullet\\ \bullet},\ \dots\}$$

The reason we introduced it in terms of strands and elaborate rules is to stress that the category can be defined without mention of functions. (In Section 3.6.20 we will drop condition (i) and then we can no longer think in terms of functions.) Also, if the reader thinks this looks a little bit like some sort of cobordism category then one goal of the drawings is achieved – otherwise perhaps this sentence can help...

There is another reason why this graphical version is purer than the algebraic. We are still talking about finite sets but now the elements have not been given names explicitly! We only distinguish them by their position in the ordering. Of course the ordering is secretly a numbering – a bijection to one of the sets $\{0, 1, 2, \dots, n\}$ – but in a sense the notion of relative position is more fundamental and in any case more flexible than the viewpoint where all elements have a fixed name.

In particular, when it comes to describing the monoidal structure, no explicit renaming is required – the principle of position takes care automatically: given two columns of dots we can simply arrange them one on top of the other:

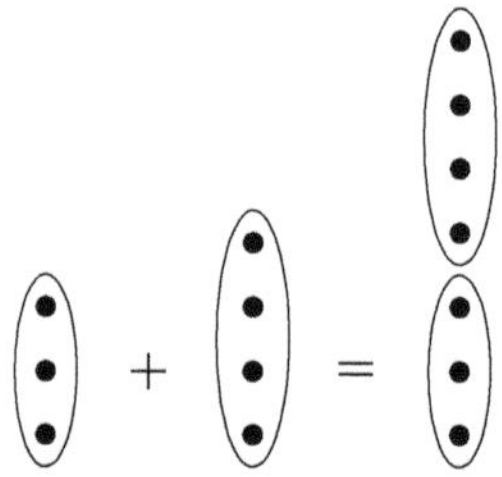

$+$ $=$

this is really stone-age mathematics!

The disjoint union of arrows is just as simple – simply place the strands side by side (i.e. on top of each other), in parallel

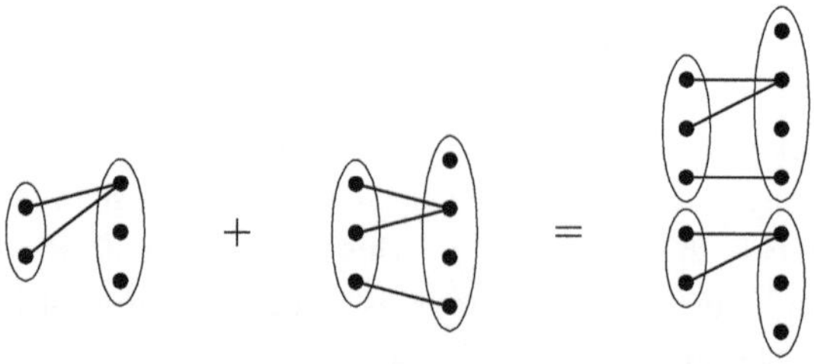

$+$ $=$

This description is simpler than the description in terms of sets and elements with names (recall the formula for the disjoint union of two order-preserving functions!).

(The above arguments are not meant to say that there is anything bad about the algebraic version – in fact, whenever you want to write anything down explicitly you need the names. But conceptually, the graphical viewpoint is rewarding.)

3.4.9 Δ is not symmetric. Symmetry would mean that for every pair of objects $\mathbf{m}, \mathbf{n}$ we should have an order-preserving map $\mathbf{m}+\mathbf{n} \to \mathbf{n}+\mathbf{m}$, and these maps should be natural with respect to other order-preserving maps, in the precise sense of 3.2.34. Now these two ordered sets are equal, so the only order-preserving isomorphism is the identity. But the identities are *not* natural in this sense. Here is an easy example to see what goes wrong: consider the identity map on $\mathbf{3}$, the unique candidate for being the symmetry $\mathbf{2}+\mathbf{1} \to \mathbf{1}+\mathbf{2}$. Let us check naturality with respect to the maps $\mu : \mathbf{2} \to \mathbf{1}$ and $\mathrm{id}_1 : \mathbf{1} \to \mathbf{1}$. The diagram does *not* commute:

$$\begin{array}{ccc} \mathbf{2}+\mathbf{1} & \xrightarrow{\mathrm{id}_3} & \mathbf{1}+\mathbf{2} \\ {\scriptstyle \mu+\mathrm{id}_1}\downarrow & & \downarrow{\scriptstyle \mathrm{id}_1+\mu} \\ \mathbf{1}+\mathbf{1} & \xrightarrow[\mathrm{id}_2]{} & \mathbf{1}+\mathbf{1} \end{array}$$

Check out the drawing:

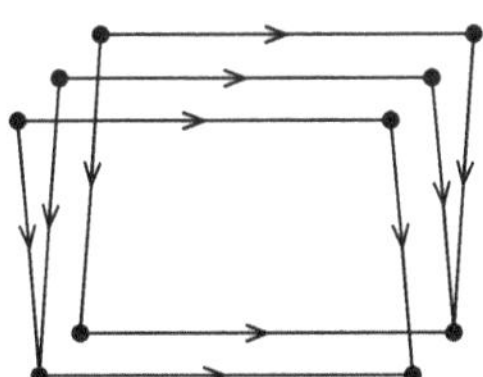

Generators and relations for Δ

The notion of generators and relations for a monoidal category was given in 1.4.

The first observation is that $\mathbf{1}$ is terminal object in Δ. This is clear from the description in terms of functions: for every $n \in \mathbb{N}$ there is exactly one function $\mathbf{n} \to \mathbf{1}$ which we denote

$$\mu^{(n)} : \mathbf{n} \to \mathbf{1}.$$

3.4.10 Generators for Δ. *The monoidal category* $(\Delta, +, \mathbf{0})$ *is generated (monoidally) by* $\mu^{(0)} : \mathbf{0} \to \mathbf{1}$ *and* $\mu^{(2)} : \mathbf{2} \to \mathbf{1}$. We draw these maps (and the identity map on $\mathbf{1}$) like this:

$\mu^{(0)}$ $\qquad$ $\mu^{(1)} = \mathrm{id}_{\mathbf{1}}$ $\qquad$ $\mu^{(2)}$

Note that we draw a little line sticking into the dot to indicate that there is a function coming in from the empty set... this is just for typographical reasons since otherwise the empty set has a natural tendency to disappear in drawings!

Proof. The first observation is that every arrow $\mathbf{m} \to \mathbf{n}$ in Δ is the sum of n arrows to $\mathbf{1}$. Indeed, we can look at the graphical representation: each dot in the target column has an inverse image, so this splits up the graph into its connected components. Since the strands are not allowed to cross over, this partition is in fact a disjoint union of arrows, in the precise sense of the monoidal operation $+$ in the category Δ. For example, for the map $f : \mathbf{7} \to \mathbf{6}$:

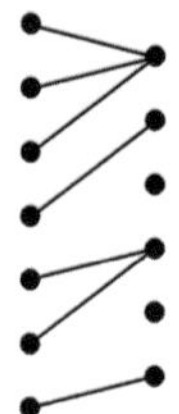

we have

$$f = \mu^{(1)} + \mu^{(0)} + \mu^{(2)} + \mu^{(0)} + \mu^{(1)} + \mu^{(3)}.$$

So some of these maps are just $\mu^{(0)} : \mathbf{0} \to \mathbf{1}$, the inclusion of the empty set in the one-element set; some are just identities $\mu^{(1)} : \mathbf{1} \to \mathbf{1}$, and the rest are maps $\mu^{(n)} : \mathbf{n} \to \mathbf{1}$ that take $n \geq 2$ elements to the same image.

Now we claim that for every $n \geq 2$ the map $\mu^{(n)} : \mathbf{n} \to \mathbf{1}$ can be obtained as a composition of maps obtained from $\mu^{(1)} = \mathrm{id} : \mathbf{1} \to \mathbf{1}$ and $\mu^{(2)} = \mu : \mathbf{2} \to \mathbf{1}$ under ordinal sum. Indeed, each such map is a collection of strands that join at the target. But we might as well let some of them join halfway to the target – this does not change the function it defines:

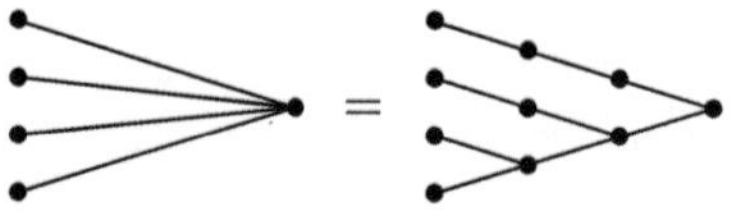

You can easily formalise this argument to prove the claim. Alternatively we could just start writing compositions involving μ and id. We easily see that we can produce maps $\mathbf{n} \to \mathbf{1}$ for every $n \geq 2$, and we have already observed that there is exactly one such map for each $n \in \mathbb{N}$. □

3.4.11 Relations. First we have the identity relations:

These are not really relations for Δ: they just express the property of the identity arrow, which holds in *any* category.

Then there are relations between the nullary operation and the binary one:

And finally the associativity relation

(You will be asked in Exercise 1 to prove that these relations suffice.)

3.4.12 Example. Here are the four possible maps $\mathbf{3} \to \mathbf{2}$.

The first two are surjective; the last two factor through $\mathbf{1}$ like this:

Concerning the map $\mathbf{3} \to \mathbf{1}$, it factors in two ways, namely the two ways given in the associativity equation above.

3.4.13 Face and degeneracy maps. It is easy to see that every arrow in Δ factors (in Δ) as a surjection followed by an injection. The surjections and injections in turn can be described quite explicitly in terms of *degeneracy maps* and *face maps*.

For each $n \in \mathbb{N}$, there are exactly $n + 1$ surjective maps called *degeneracy maps* σ_k^n $(k = 0, \ldots, n)$:

$$\sigma_k^n : \mathbf{n+2} \longrightarrow \mathbf{n+1}$$

$$i \longmapsto \begin{cases} i & i \leq k \\ i-1 & i > k. \end{cases}$$

In other words, the element $k \in \mathbf{n+1}$ is hit twice by σ_k^n. Here is a picture of σ_2^3:

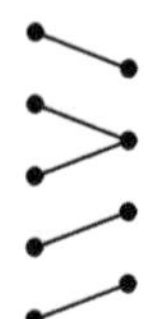

Clearly we have

$$\sigma_k^n = \mathrm{id}_{\mathbf{k}} + \mu + \mathrm{id}_{\mathbf{n-k}},$$

and in particular $\sigma_0^0 = \mu$. (The indexation choice – superscript n for a map $\mathbf{n+2} \to \mathbf{n+1}$ – was made to obtain a neat formula.) Now it is not difficult to prove that every (nonidentity) surjection in Δ can be written as a composition of face maps.

In the other direction, for fixed $n \in \mathbb{N}$ there are $n+1$ injective maps δ_k^n $(k = 0, \ldots, n)$ called *face maps*,

$$\begin{aligned} \delta_k^n : \mathbf{n} &\longrightarrow \mathbf{n+1} \\ i &\longmapsto \begin{cases} i & i < k \\ i+1 & i \geq k. \end{cases} \end{aligned}$$

In other words, δ_k^n is the injection that fails to hit $k \in \mathbf{n+1}$. Here is a picture of δ_2^3:

Clearly we have

$$\delta_k^n = \mathrm{id}_{\mathbf{k}} + \eta + \mathrm{id}_{\mathbf{n-k}},$$

and in particular $\delta_0^0 = \eta$. Every (nonidentity) injection can be written as a composition of face maps.

So, in conclusion, the collection of all face and degeneracy maps generates Δ as a category.

The relations we found above then translate into the *(co)simplicial identities* (where we suppress the upper indices):

$$\begin{aligned} \delta_j \delta_i &= \delta_i \delta_{j+1} & i &\leq j, \\ \sigma_i \sigma_j &= \sigma_{j+1} \sigma_i & i &\leq j, \\ \delta_i \sigma_j &= \begin{cases} \sigma_{j-1} \delta_i & i < j, \\ \mathrm{id} & i = j, i = j+1, \\ \sigma_j \delta_{i-1} & i > j+1. \end{cases} \end{aligned} \tag{3.4.14}$$

The relations involving only face maps are more or less trivial: they essentially amount to shuffling around identity maps. The relations involving only degeneracy maps follow from the associativity relation (the relation $\sigma_0^1 \sigma_0^0 = \sigma_1^1 \sigma_0^0$ is precisely the associativity relation; the others are variations) and the identity argument again. Finally, the mixed relations are essentially the unit relations.

This is a presentation of Δ in terms of generators and relations – not as a monoidal category but plainly as a category. Note that there are infinitely many generators and infinitely many relations, but that they come in series: for this reason we can grasp them even though they are so many. The reason why they come in series is of course the monoidal structure, which in some sense gives a much clearer picture.

3.4.15 Aside: the topologist's delta. The simplex category has another variant which is important in topology, and which is also usually denoted Δ. Here we denote it $\triangle$ just to be able to distinguish it from our Δ. It is the full subcategory of our Δ consisting of the positive ordinals. Traditionally these ordinals are then denoted by one less than we have done here, and often they are written with square brackets, i.e. one puts

$$[0] = \{0\}, \quad [1] = \{0, 1\}, \quad [2] = \{0, 1, 2\}, \ldots$$

This convention is practical because the naming number refers to the dimension of the corresponding simplex: $[0]$ is the category $\bullet$ (dimension 0), $[1]$ is the category $\bullet \to \bullet$ (dimension 1), and so on.

A *simplicial set* is a functor $\triangle^{\text{op}} \to \textbf{\textit{Set}}$. It amounts to having a sequence of sets and set maps

$$X_0 \;\substack{\longleftarrow \\ \longrightarrow \\ \longleftarrow}\; X_1 \;\substack{\longleftarrow \\ \longrightarrow \\ \longleftarrow \\ \longrightarrow \\ \longleftarrow}\; X_2 \qquad \cdots$$

which satisfy identities dual to 3.4.14. The elements of X_k are called k-simplices. To every simplicial set one can associate a topological space called the geometric realisation, roughly by gluing together simplices along faces they have in common according to the combinatorics of the diagram. Conversely, to a topological space one can associate a simplicial set (the singular complex). These correspondences constitute a very close relation between simplicial sets and spaces, and most of homotopy theory can be carried out in the context of simplicial sets (see for example Goerss and Jardine [24]).

Simplicial sets are also very important in category theory. Every (small) category gives rise to a simplicial set called its *nerve*, denoted $N : \triangle^{\text{op}} \to \textbf{\textit{Set}}$. Namely, the value of N on $[0]$ is the set of objects; $[1]$ is sent to the set of all arrows; $[n]$ is sent to the set of n-tuples of composable arrows. Composing two consecutive arrows in such a chain defines the face maps, while inserting an identity arrow somewhere in the chain defines the degeneracy maps. These maps satisfy the simplicial identities, so N is a simplicial set.

In this way, the topologist's delta is a sort of bridge between category theory and topology. In these contexts the empty ordinal (empty simplex) is not used, but in our context it is important to keep it, because without it we could not have a monoidal structure.

The symmetric equivalent: finite cardinals

We now come to the symmetric case. From the graphical viewpoint what we do (compared to the construction of Δ) is simply to allow 'crossing-over'. In terms of finite sets it means there is no longer any order relation in play; there is no hierarchy among the elements in the set, so everything is symmetric or homogeneous...

The exposition follows closely the nonsymmetric case – in fact we have already done a lot of the work.

3.4.16 The category of finite sets. Let ***FinSet*** denote the category whose objects are the finite sets, and whose arrows are all maps between finite sets (so it is a full subcategory of ***Set***.) Sometimes a set can happen to possess an order, but regarded as an object in ***FinSet*** we simply ignore that order; in particular, given two such sets which happen to have an order, the maps between them are not required to respect the order. For the record, let us draw this diagram of categories and functors:

$$\begin{array}{ccc} \textbf{\textit{OrdSet}} & \longrightarrow & \textbf{\textit{Set}} \\ \cup & & \cup \\ \textbf{\textit{FinOrd}} & \longrightarrow & \textbf{\textit{FinSet}} \end{array}$$

The upwards arrows $\cup$ are full embeddings/inclusions, the rightwards arrows are the forgetful functors.

Consider now the monoidal structure on ***Set*** given by disjoint union and empty set (cf. 1.3.24). The disjoint union of two finite sets is again finite, and also the empty set is finite, so there is induced a monoidal structure on ***FinSet***. The two categories on the left in the diagrams are also monoidal under disjoint union, and clearly the four arrows in the diagram are monoidal functors. We are mostly interested in the forgetful monoidal functor

$$(\textbf{\textit{FinOrd}}, \coprod, \varnothing) \longrightarrow (\textbf{\textit{FinSet}}, \coprod, \varnothing).$$

3.4.17 Symmetry. An important property of ***FinSet*** (which is not shared by ***FinOrd***) is that *disjoint union is the coproduct in* ***FinSet***, just as it is in ***Set***, cf. 1.3.24.

Hence (cf. 3.2.42) there is a unique symmetric structure τ on $(\textbf{\textit{FinSet}}, \coprod, \varnothing)$, which is just the one induced from the symmetry on ***Set*** – mere interchange of 'factors'. So $(\textbf{\textit{FinSet}}, \coprod, \varnothing, \tau)$ *is a symmetric monoidal category.*

3.4.18 Finite cardinals. The category Φ of finite cardinals is defined as a skeleton of ***FinSet***. Since two finite sets are isomorphic if and only if they have the same number of elements, there will be one object in Φ for each $n \in \mathbb{N}$. Let $\mathbf{n}$ denote the set $\{0, 1, 2, \ldots, n-1\}$, then Φ is the full subcategory of ***FinSet*** given by

$$\Phi = \{\mathbf{0}, \mathbf{1}, \mathbf{2}, \ldots\}.$$

The objects of Φ are called *finite cardinals*. In 3.4.4 we used the symbol $\mathbf{n}$ to denote the *ordered* set $\{0, 1, 2, \ldots, n-1\}$, which we called an *ordinal* (object in Δ). As *sets* they are the same (i.e. forgetting the ordering), so we allow ourselves to say that Δ and Φ have the same objects. A more precise way of saying this is that the forgetful functor ***FinOrd*** $\to$ ***FinSet*** restricts to an embedding of Δ into Φ:

$$\begin{array}{ccc} \textbf{\textit{FinOrd}} & \longrightarrow & \textbf{\textit{FinSet}} \\ \cup & & \cup \\ \Delta & \hookrightarrow & \Phi \end{array}$$

which is furthermore a bijection on objects. However, Φ clearly has many more arrows than Δ, since they are not required to respect any order.

Cardinal sum (which to two cardinals $\mathbf{m}$ and $\mathbf{n}$ associates the cardinal corresponding to the natural number sum $m + n$) makes Φ into a monoidal category – clearly $\mathbf{0}$ is the neutral object. It coincides with ordinal sum, and the formulae for inclusion of the sets $\mathbf{m}$, $\mathbf{n}$ into their disjoint union $\mathbf{m} + \mathbf{n}$ (as well as the formulae for the cardinal sum of two functions) are the same as the formulae in Δ. (This amounts to saying that the embedding

$$(\Delta, +, \mathbf{0}) \hookrightarrow (\Phi, +, \mathbf{0})$$

is monoidal.)

It is easy to adapt the arguments for ***FinSet*** to show that

3.4.19 Lemma. $+$ *is the coproduct in* Φ.

And consequently, *there is a unique symmetric structure on* $(\Phi, +, \mathbf{0})$.

3.4.20 Remark. The category Φ also has products: the *cardinal product* which to two cardinals $\mathbf{m}$ and $\mathbf{n}$ associates the cardinal corresponding to the natural number product mn, is the categorical product. In particular, the object $\mathbf{1}$ is terminal object – clearly for each $n \in \mathbb{N}$ there is exactly one map $\mathbf{n} \to \mathbf{1}$.

The category Φ is is the *categorification* of $\mathbb{N}$, in the sense that all important set-theoretic properties of $\mathbb{N}$ are in fact reflections of category-theoretic properties of Φ. We have already seen how cardinal sum and product correspond to the usual sum and product in $\mathbb{N}$. For a pleasant introduction to the philosophy of categorification, see Baez and Dolan [10].

3.4.21 Categorical viewpoint. In 3.4.5 we characterised the ordinal $\mathbf{n}$ as the category

$$\{0 \to 1 \to 2 \to \cdots \to n-1\}$$

(the category contains also all the compositions of these arrows, so its graph is an oriented simplex). Then we characterised Δ as the full subcategory of ***Cat*** given by

$$\{\mathbf{0}, \mathbf{1}, \mathbf{2}, \dots\}.$$

We might ask for a similar category interpretation of cardinals, such that Φ could be characterised as the full subcategory of ***Cat*** consisting of those categories. This is easy: each set $\{0, 1, \dots, n-1\}$ determines a discrete category (see 3.2.20), which we denote by $\mathbf{n}_0$. It looks like this:

$$\{0 \quad 1 \quad 2 \quad \cdots \quad n-1\}$$

(there are no arrows other than the identities). Now Φ is the full subcategory of ***Cat*** determined by these discrete categories. Indeed, a functor on a discrete category is determined completely by what it does on objects, because the only arrows are the identity arrows, and they are bound to go to the identity arrows of the image object. This is just to say that a functor between discrete categories amounts to a function on the underlying sets. So every functor $\mathbf{m} \to \mathbf{n}$ induces a functor $\mathbf{m}_0 \to \mathbf{n}_0$, and this describes the embedding $\Delta \hookrightarrow \Phi$.

Curiously there is another way of describing Φ as a subcategory of ***Cat***, which is opposite in spirit. Let $\overline{\mathbf{n}}$ be the category with n objects and a unique (invertible) arrow between any two objects – it looks like this:

$$\{0 \leftrightarrow 1 \leftrightarrow 2 \leftrightarrow \cdots \leftrightarrow n-1\}$$

(the category contains also all the compositions of these arrows). This defines an equivalence relation on $\overline{\mathbf{n}}$ namely the one usually called the *chaotic equivalence relation*, where everybody is related to everybody. In terms of graphs we have the complete graph on $\{0, 1, \ldots, n-1\}$, or an unoriented n-simplex.

Now what are the functors between such categories? They are determined completely as soon as their values on the objects are specified, because once the images of two objects are fixed then the unique arrow between those two objects must map to the unique arrow between the images.

So now we might as well characterise Φ as the full subcategory of ***Cat*** given by

$$\{\overline{\mathbf{0}}, \overline{\mathbf{1}}, \overline{\mathbf{2}}, \ldots\}.$$

Every functor $\mathbf{m} \to \mathbf{n}$ induces a functor $\overline{\mathbf{m}} \to \overline{\mathbf{n}}$, and this describes the embedding $\Delta \hookrightarrow \Phi$.

(The moral is that an equivalence relation in which everybody is related contains exactly the same information as if nobody were related, cf. also the fairy tale *The Tinder-Box* [4] by H. C. Andersen.)

This last description of Φ as subcategory in ***Cat*** is given only in order to justify calling Φ a sort of simplex category. The first description is actually better. For example the embedding $(\Phi, +, \mathbf{0}) \hookrightarrow (\boldsymbol{Cat}, \coprod, \emptyset)$ defined in terms of the discrete description is monoidal, while the embedding given by the second description is not. (To see this, note first that the coproduct of two categories is given by taking disjoint union on the set of objects and disjoint union on the set of arrows. So the coproduct of the discrete categories $\mathbf{m}_0$ and $\mathbf{n}_0$ is precisely $(\mathbf{m}+\mathbf{n})_0$, which shows that the first embedding is monoidal. In contrast, $\overline{\mathbf{m}}\coprod\overline{\mathbf{n}}$ is a nonconnected category (unless one of the summands is $\overline{\mathbf{0}}$), while $\overline{\mathbf{m}+\mathbf{n}}$ is connected, so the second embedding is not monoidal.)

3.4.22 Graphical description of Φ. Since we are now talking about sets without any order, we ought to picture the elements without any order, something like this:

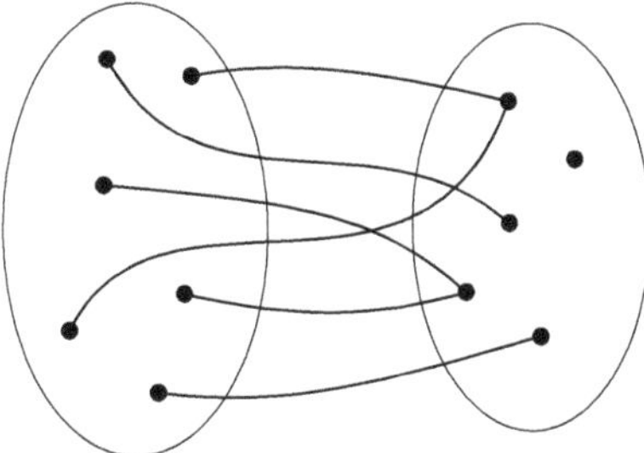

but then we would have to name the elements explicitly to keep track of which are which. Instead, since anyway the cardinals coincide with the ordinals, and since the latter come with internal order, we might as well use this order to line up their elements in a column. But we do *not* require the maps to preserve that order, so our maps will rather look like this:

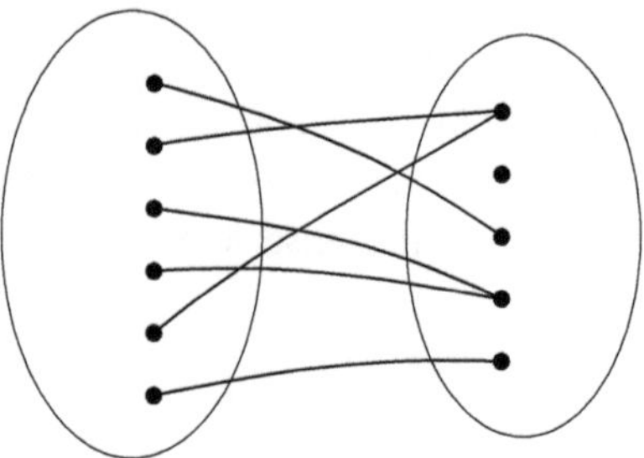

Here is the definition in graphical terms: the *objects* are finite sets of dots arranged in a column like this

An *arrow* from one column to another is a collection of strands starting at the dots in the source column and ending at dots in the target column, subject to the following single rule:

(i) for each dot in the source column there is exactly one strand coming out (and going to the target column)

Compared to the graphical definition of Δ, the difference is that we now allow the strands to cross over each other. For this reason there is no need of fine print conditions that complicated the description of Δ: now the strands may emanate in any direction they want, and the picture does not necessarily have to be plane.

The pictures of composition and disjoint union of maps are described just as for Δ.

Generators and relations for Φ

3.4.23 The category of finite sets and bijections. Let $\boldsymbol{FinSet}_0$ denote the category whose objects are the finite sets, and whose arrows are the bijections between finite sets. So this category has the same objects as ***FinSet*** but less arrows.

Let Σ be the skeleton of $\boldsymbol{FinSet}_0$ defined by

$$\{\mathbf{0}, \mathbf{1}, \mathbf{2}, \ldots\}.$$

It is the subcategory of Φ consisting of all the objects but only the invertible arrows. Since there are no bijections between sets of different cardinality, the graph of the category Σ is disconnected, with a connected component for each object $\mathbf{n}$. In other words Σ is the disjoint union of the monoids (actually groups) $\mathrm{End}_\Sigma(\mathbf{n})$ – recall from 3.1.19 the correspondence between monoids and 1-object categories. Which are the arrows from $\mathbf{n}$ to $\mathbf{n}$? Well, they are precisely the permutations of the elements in the set $\mathbf{n} = \{0, 1, \ldots, n-1\}$. So $\mathrm{End}_\Sigma(\mathbf{n}) = \mathfrak{S}_n$, the symmetric group on n letters. So Σ is the disjoint union of all those 1-object categories:

$$\Sigma = \coprod_{k=0}^{\infty} \mathfrak{S}_k$$

(where we define $\mathfrak{S}_0$ to be the trivial group).

Now $(\Sigma, +, \mathbf{0})$ is a monoidal category (subcategory of Φ), since the sum of two bijections is again a bijection. It is generated monoidally by the transposition

(recall that each symmetric group $\mathfrak{S}_k$ ($k \geq 2$) is generated by transpositions, cf. 1.4.2).

So now we have two subcategories in Φ: Δ and Σ...

3.4.24 Lemma. *Every arrow in Φ can be factored as a permutation followed by an order-preserving map. In other words,*

$$\Phi = \Sigma\Delta.$$

Proof. From the graphical viewpoint this is clear: take any arrow and factor it by taking a vertical cut so far to the right that all the cross-overs occur on the left-hand side of it:

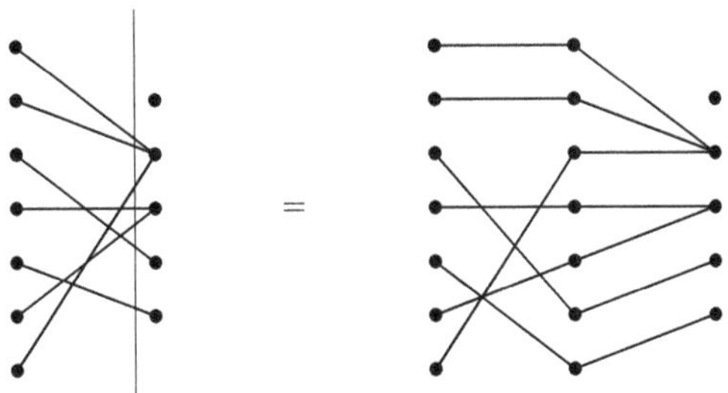

□

Note that it is not true that we could factor every map as first an order-preserving map and then a permutation of the target! The problem is that a dot in the target might be hit by various strands, and clearly this cannot happen with a permutation which by definition is one-to-one. Here is a simple example:

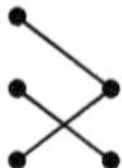

Observe also that the factorisation is not unique. The order-preserving part f *is* unique, but for the permutation part, we have the freedom to permute those elements which have the same image under f.

3.4.25 Generators for Φ. It follows immediately from the lemma that this is a complete set of generators:

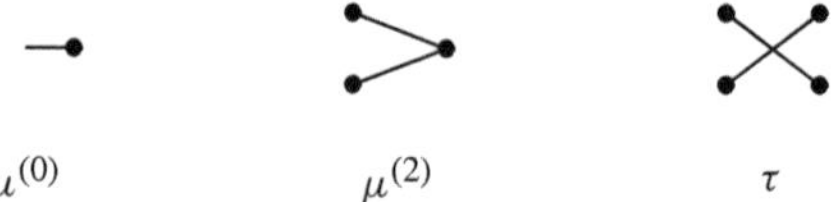

$\mu^{(0)}$ $\mu^{(2)}$ τ

3.4.26 Relations. Of course there are those relations coming from Δ (cf. 3.4.11):

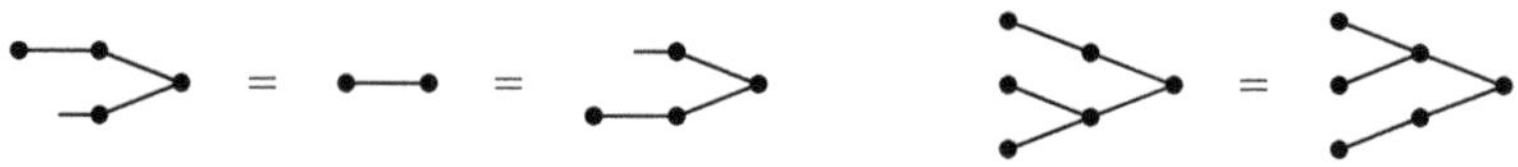

and those relations coming from Σ (the symmetric group relations, see 1.4.2)

Next, since Φ is a symmetric monoidal category we have the relations expressing the naturality of the twist map (cf. 3.2.36). By the arguments given in 1.4.35 it is enough to consider naturality with respect to the 'generators parallel with identities'. These relations are:

Finally there is one more relation, namely the commutativity relation

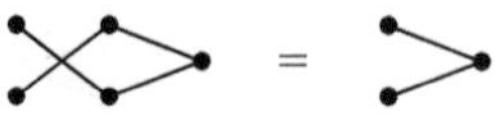

valid since **1** is a terminal object. More generally, in combination with the associativity relation, this implies that it has no effect to permute the input dots within a 'connected component', i.e. dots which have the same image.

To see that these relations suffice, we must show that every composition of the generators can be brought on the form of the previous lemma. (This form is not unique, but it differs only by some permutation on the input side, within the connected components (inverse images of points) and our last relation (the commutativity relation) accounts for that.) So given an expression in the generators we need to move the twist maps to the left, until they come before any of the order-preserving maps. But the naturality relations allow us precisely to do that.

Exercises

1. Show that the relations listed in 3.4.11 suffice. (Hint: define a 'normal form' as in the end of 1.4.)
2. Show that there are $\binom{m+n-1}{n-1}$ different arrows from $\mathbf{m}$ to $\mathbf{n}$ in Δ. (Hint: giving such an arrow amounts to dividing the m input dots into n parts, according to their image.)
3. Show that the injective maps in Δ are precisely its monomorphisms. Let Δ_{mono} denote the subcategory of Δ having the same objects but only the monomorphisms as arrows. Describe Δ_{mono} in terms of generators and relations.
4. Show that the surjective maps in Δ are precisely its epimorphisms. Let Δ_{epi} denote the subcategory of Δ having the same objects but only the epimorphisms as arrows. Describe Δ_{epi} in terms of generators and relations.
5. Draw pictures of the symmetry map $\sigma : \mathbf{2}+\mathbf{5} \to \mathbf{5}+\mathbf{2}$ in Φ. Then draw pictures of the symmetry map $\mathbf{3}+\mathbf{4} \to \mathbf{4}+\mathbf{3}$. Write the maps in terms of the generator $\times$. Write down a general formula for the symmetry maps in terms of generators.
6. *Artin's braid group* on k letters, denoted $\mathfrak{B}_k$, is a lot like the symmetric group $\mathfrak{S}_k$, but instead of merely recording which letters change place we now keep track of *how*: instead of just displaying two columns of dots and matching the dots by drawing a connecting strand, the important data now is the topology of the strands themselves, how they wrap around each other and who goes under and who goes over – for this to make sense, the whole picture must be embedded in 3-space. It is not difficult to grasp this group geometrically by looking at a couple of pictures. Here is an arbitrary snapshot from $\mathfrak{B}_4$:

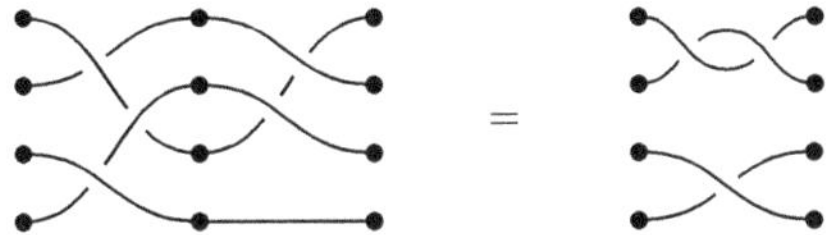

And here is an important picture from $\mathfrak{B}_3$:

The easiest description is in terms of generators and relations (compare 1.4.2). The generators for $\mathfrak{B}_k$ are the symbols $\tau_1 \ldots, \tau_{k-1}$, and the relations are

$$\begin{aligned} \tau_i \tau_j \tau_i &= \tau_j \tau_i \tau_j \qquad \text{for } j = i + 1 \\ \tau_i \tau_j &= \tau_j \tau_i \qquad \text{for } j > i + 1. \end{aligned}$$

(Note that according to our convention for listing generators and relations for a group (1.4.1) we have not listed the inverses $\tau_1^{-1}, \ldots, \tau_{k-1}^{-1}$; they are assumed to exist too, as soon as we are speaking about generating a *group*, not just a monoid.)

(i) The first point in this exercise is to describe carefully the connection between the abstract description in terms of generators and relations, and the graphical description – in fact, you have to provide the graphical description yourself, based on the few drawings and the abstract description!

(ii) Recognise $\mathfrak{B}_1$ and $\mathfrak{B}_2$ as well known groups.

(iii) Define a natural group homomorphism $\mathfrak{B}_k \to \mathfrak{S}_k$.

Now denote by B the disjoint union of all the groups $\mathfrak{B}_k$, $k \geq 0$, in the same way as Σ was defined in terms of the symmetric groups in 3.4.23. It is called the *braid category*. Still mimicking 3.4.23, give B monoidal structure, denoted $(\mathrm{B}, +, \mathbf{0})$, in such a way that $\mathrm{B} \to \Sigma$ becomes a monoidal functor. Describe $(\mathrm{B}, +, \mathbf{0})$ in terms of generators and relations as a monoidal category. (The B is meant as a capital beta, in analogy with Δ, Φ, and Σ.)

(iv) The notions of braiding and braided monoidal category were given on page 169. Consider the maps $\mathbf{m} + \mathbf{n} \to \mathbf{n} + \mathbf{m}$ consisting in crossing the n upper strands over the m lower strands. Here is a drawing of $\mathbf{2} + \mathbf{3} \to \mathbf{3} + \mathbf{2}$:

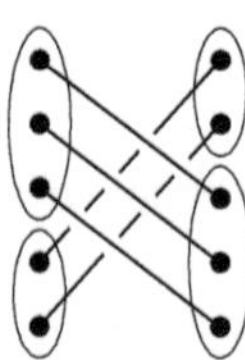

Show that this family of maps defines a braiding on B.

(v) Define a notion of *braided monoidal functor* (do not be too creative!) and show that $\mathrm{B} \to \Sigma$ is such a functor.

3.5 Monoids in monoidal categories, and monoidal functors from $\mathbf{\Delta}$

Monoids in monoidal categories

Monoidal categories are our context for defining monoids! Then expressions of type $X \square Y$ play the rôle of $A \times B$ in the category of sets. So while formally $\square$ is a map that to a pair of objects (X, Y) associates a single object denoted $X \square Y$, what will happen in the following is that we abusively think of $X \square Y$ as a sort of product, as if it were made up of pairs. This abuse is common for example with vector spaces, where one often thinks about an element of $V \otimes W$ as a pair consisting of one element from V and another from W, even when we know that in reality a vector in the tensor product space is not in general of this form, but only a linear combination of such vectors.

3.5.1 Definition of monoid (in an arbitrary monoidal category). Let $(\mathbf{V}, \square, I)$ be a monoidal category. A monoid in $\mathbf{V}$ is an object M together with two arrows

$$\mu : M \square M \to M, \qquad \eta : I \to M$$

such that these three diagrams commute:

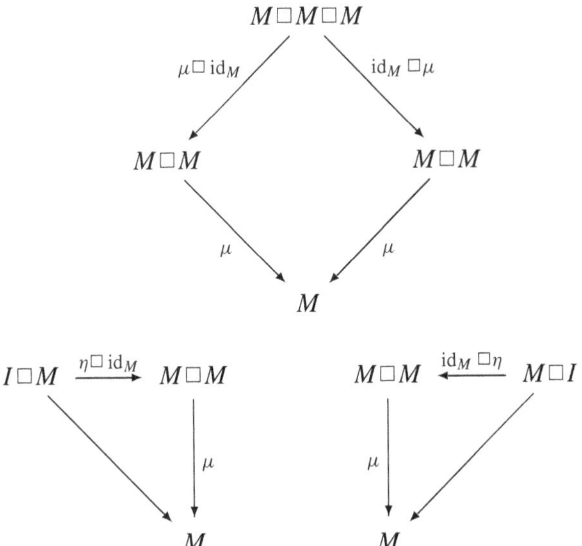

The diagonal maps without labels are the identifications of 3.2.8.

It is often convenient to picture M itself as single dot, and let the two structure maps be pictured like this:

μ η

Then the axioms read

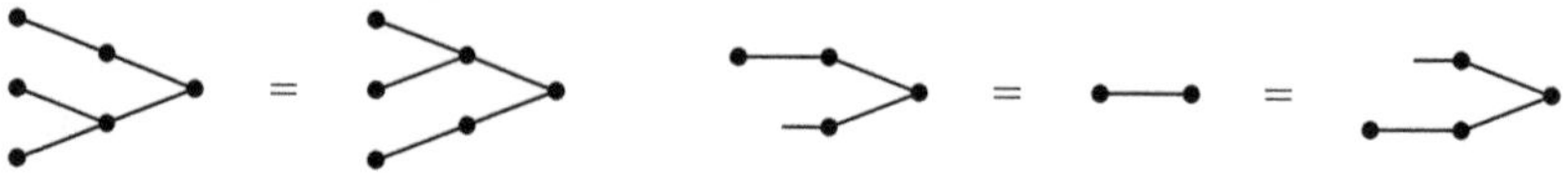

3.5.2 Monoid homomorphisms. A *monoid homomorphism* in $\boldsymbol{V}$ between two monoids (M, μ, η) and (M', μ', η') is an arrow $\phi : M \to M'$ that commutes with all the monoid structure. Precisely,

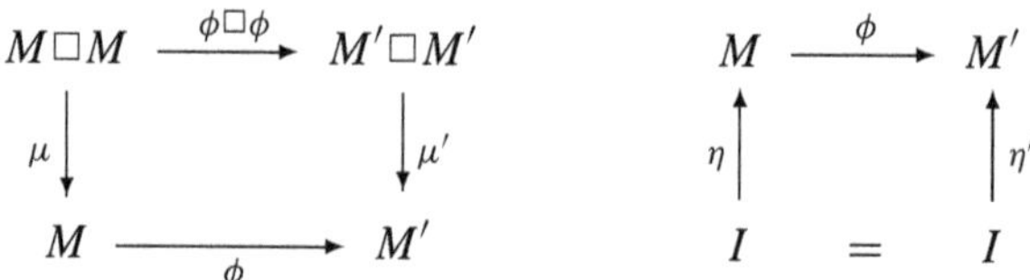

3.5.3 The category of monoids. One easily checks that the composition of two monoid homomorphisms is again a monoid homomorphism, and that the identity arrow is a monoid homomorphism, so altogether: there is a category denoted $\boldsymbol{Mon}(\boldsymbol{V})$ whose objects are the monoids in $\boldsymbol{V}$ and whose arrows are the monoid homomorphisms in $\boldsymbol{V}$. A monoid homomorphism is called an *isomorphism of monoids* if there exists a two-sided inverse which is also a monoid homomorphism.

3.5.4 Remarks on the 'neutral arrow'. Since in these notes we put a lot of effort in distinguishing structures from properties, we should mention that the neutral arrow of a monoid, which we have introduced as a structure, can also be seen as a property: if a neutral element exists then it is uniquely determined. This remark was made in the context of set monoids in Exercise 3 on page 148.

In the general context of monoids in a monoidal category the same remark holds: two arrows $I \to M$ which satisfy the neutral axiom must necessarily coincide. The idea of the proof is the same as in the case of set monoids, but since we do not have elements at our disposal, we are forced to write it out in terms of arrows and beautiful diagrams like this: suppose $\eta : I \to M$ and

$\eta' : I \to M$ both satisfy the neutral axiom, then

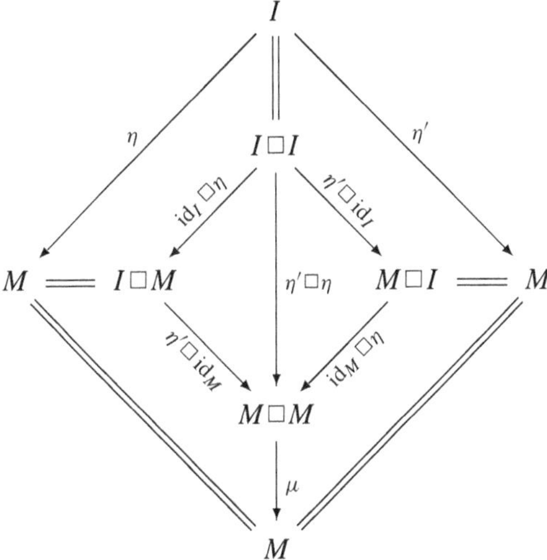

All the inner cells of this diagram commute by the unit axioms (of η and η'), so we conclude that also the outer square commutes. Thus $\eta = \eta'$.

One reason for preferring to regard the neutral as a structure appears when we consider arrows. As a general principle, in a category whose objects are sets with structure, the arrows should be the set maps that preserve the structure. In the case of monoids it is desirable that neutral elements are preserved, so we had better stipulate that neutral element be a structure. (Note that there exist semi-monoid homomorphisms between monoids that are not monoid homomorphisms, cf. Exercise 4 on page 148.)

3.5.5 Lemma. If $(\boldsymbol{V}, \otimes, I, \tau)$ is a symmetric monoidal category, then $\boldsymbol{Mon}(\boldsymbol{V})$ is canonically a monoidal category.

Proof. Let (M, μ, η) and (M', μ', η') be two monoids in $\boldsymbol{V}$. Define a new monoid $(M \Box M', \mu \Box \mu', \eta \Box \eta')$ by setting $M \Box M' := M \otimes M'$ and $\eta \Box \eta' := \eta \otimes \eta'$, and define the new multiplication $\mu \Box \mu'$ by

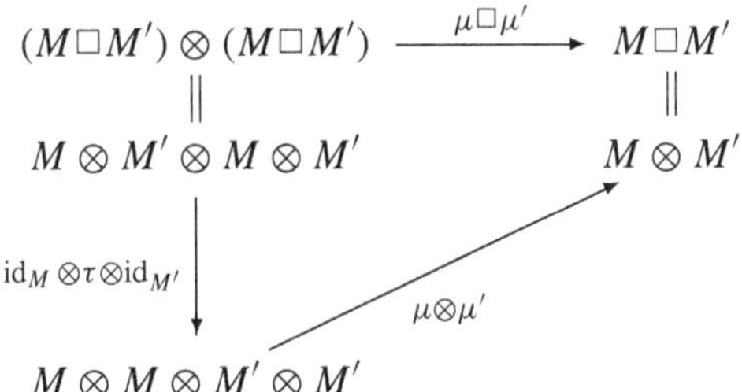

Note that the twist map is involved, so it is crucial that we are in a *symmetric* monoidal category. Here is the drawing for the new multiplication:

It is easy to check that the unit axioms hold. The associativity equation reads

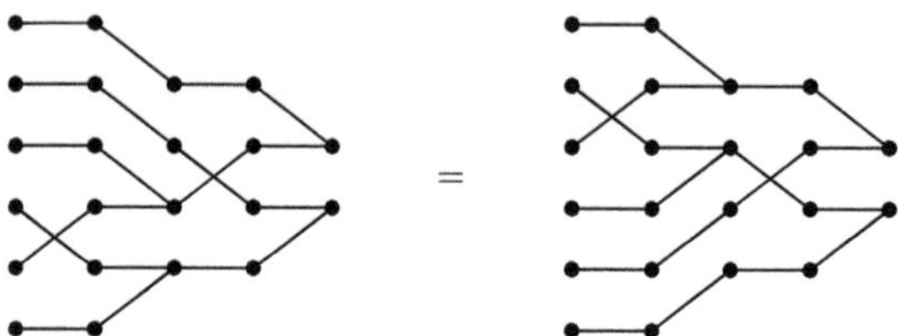

We already proved this equation in another setting. That was in 2.4.6 when we proved that the tensor product of two algebras is again an algebra. For the time being we can just copy over the proof and be happy that it works. In 3.5.11 we will see the explanation, namely that $\Bbbk$-algebras are just monoids in $(\boldsymbol{Vect}_{\Bbbk}, \otimes, \Bbbk, \sigma)$.

(Working out what the definition of $\square$ should be on arrows is left as an exercise.) $\square$

3.5.6 Commutative monoids. Now suppose $(\boldsymbol{V}, \square, I, \tau)$ is a *symmetric* monoidal category. Intuitively, a monoid $(M, ., 1)$ in $\boldsymbol{V}$ is called commutative if for all elements $a, b \in M$ we have $a.b = b.a$ (cf. 3.1.9). However, this might not make sense at all, because the objects of $\boldsymbol{V}$ might not be sets, and therefore it would be nonsense to talk about elements! But we can just express what we want in terms of arrows and commutative diagrams: a monoid M in $\boldsymbol{V}$ is called *commutative* if the multiplication $\mu : M \square M \to M$ is compatible with the twist map. That is, we have a commutative diagram

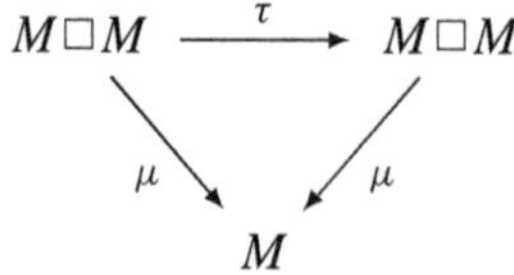

Graphically,

Examples

3.5.7 Monoids in *Set*. We have seen in Example 3.2.21 that $(\boldsymbol{Set}, \times, 1)$ is a monoidal category. A monoid herein is just a monoid in the sense of Section 3.1. So $\boldsymbol{Mon} = \boldsymbol{Mon}((\boldsymbol{Set}, \times, 1))$. The two notions of commutativity coincide.

3.5.8 Monoids in *Cat*. We noticed in 3.2.23 that $(\boldsymbol{Cat}, \times, \mathbf{1})$ is a monoidal category. A monoid in $(\boldsymbol{Cat}, \times, \mathbf{1})$ is nothing but a (strict) monoidal category.

Note that a symmetric monoidal category is not necessarily a commutative monoid in $\boldsymbol{Cat}$. Indeed, being symmetric does not mean that the multiplication is completely compatible with the background symmetry (given by the interchange of factors in the cartesian product), but merely that there is an isomorphism between them...

3.5.9 Rings. The category of abelian groups is a monoidal category under tensor product, denoted $(\mathbf{Ab}, \otimes_{\mathbb{Z}}, \mathbb{Z})$.

Now we claim that *a monoid in* $(\mathbf{Ab}, \otimes_{\mathbb{Z}}, \mathbb{Z})$ *is exactly a ring.*

Let us be detailed here: a ring is usually defined as a set R equipped with two binary operations denoted $+$ and $\cdot$, and two special elements denoted 0 and 1, subject to a long list of axioms. The axioms which pronounce themselves on the operation $+$ amount precisely to saying that $(R, +, 0)$ is an abelian group. The axioms concerning the other operation amount to the statement that $(R, \cdot, 1)$ is a monoid (in $(\boldsymbol{Set}, \times, 1)$), in other words we have maps $\mu : R \times R \to R$ and $\eta : 1 \to R$ as in Section 3.1. Finally there are a couple of axioms concerning the compatibility of the two structures $+$ and $\cdot$, namely the distributivity axioms. These axioms amount to saying that the multiplication map

$$\begin{aligned} R \times R &\longrightarrow R \\ (x, y) &\longmapsto x \cdot y \end{aligned}$$

is a group homomorphism in each variable, i.e. a bilinear map.

Now here comes the whole point: by the universal property of the tensor product, giving a bilinear map $R \times R \to R$ is the same as giving a linear map $R \otimes_{\mathbb{Z}} R \to R$, and in fact, in view of some of the axioms, this provides a monoid structure on R in the monoidal category $(\mathbf{Ab}, \otimes_{\mathbb{Z}}, \mathbb{Z})$. In detail, given the bilinear map $R \times R \to R$, the universal property gives a linear map $R \otimes_{\mathbb{Z}} R \to R$ and a factorisation of the original multiplication map through this tensor product. And conversely, given $R \otimes_{\mathbb{Z}} R \to R$, we recover the bilinear map $R \times R \to R$ by composing with the structure map $R \times R \to R \otimes_{\mathbb{Z}} R$.

To be pedantic, there is another universal property in play here: in the original definition of a ring we had a set map $\eta : 1 \to R$ as part of the monoid structure $(R, \cdot, 1)$. Now since $(R, +, 0)$ is in fact a group, and since $\mathbb{Z}$ is the free group on 1, giving a set map $1 \to R$ is the same as giving a group homomorphism $\mathbb{Z} \to R$ (which we also denote η). (We can regard this as the zeroth tensor power part of the universal property of the tensor product.)

So in conclusion, the long list of data and axioms that define a ring in terms of sets (i.e. as a structure on an object in $(\boldsymbol{Set}, \times, 1)$) can be expressed equivalently by giving a single structure on an object in $(\boldsymbol{Ab}, \otimes_{\mathbb{Z}}, \mathbb{Z})$, namely the monoid structure in this monoidal category.

In fact the correspondence between monoids-in-$\boldsymbol{Ab}$ and rings is functorial. There is also a correspondence between monoid homomorphisms in $\boldsymbol{Ab}$ and ring homomorphisms: a monoid homomorphism $A \to B$ between two monoids in $\boldsymbol{Ab}$ is a map of abelian groups which is compatible with the monoid structure maps. Altogether it is a map of sets compatible with all the structure: group structure and monoid structure on top of that. In short, it is a ring homomorphism. Altogether

$$\boldsymbol{Mon}(\boldsymbol{Ab}) = \boldsymbol{Ring}.$$

(Here ***Ring*** denotes the category of rings.)

Similarly, commutative monoids in $(\boldsymbol{Ab}, \otimes_{\mathbb{Z}}, \mathbb{Z}, \tau)$ (τ is the canonical twist map) are just commutative rings:

$$\boldsymbol{cMon}(\boldsymbol{Ab}) = \boldsymbol{cRing}$$

(the category of commutative rings).

3.5.10 Modules and algebras over a ring. Let R be a commutative ring (i.e. a commutative monoid in $(\boldsymbol{Ab}, \otimes_{\mathbb{Z}}, \mathbb{Z}, \tau)$). An R-module is an abelian group M with an R-action (see 3.1.21). If we make everything internal to the monoidal category $(\boldsymbol{Ab}, \otimes_{\mathbb{Z}}, \mathbb{Z})$ we can say that it is an object equipped with an action of the commutative monoid R. The R-modules form a monoidal category $(\boldsymbol{Mod}_R, \otimes_R, R)$.

Now we know that abelian groups can be regarded as $\mathbb{Z}$-modules, and that rings are $\mathbb{Z}$-algebras in this viewpoint. Replacing $\mathbb{Z}$ by a general commutative ring R, the arguments in the example above show that *an R-algebra is a monoid in $(\boldsymbol{Mod}_R, \otimes_R, R)$*.

3.5.11 $\Bbbk$-algebras! In particular, if R is our fixed field $\Bbbk$, then its modules are precisely the $\Bbbk$-vector spaces (cf. 2.1.1). Thus *a monoid in* $(\textbf{Vect}_\Bbbk, \otimes_\Bbbk, \Bbbk)$ *is precisely a* $\Bbbk$*-algebra.* (In fact we defined $\Bbbk$-algebras just like this in 2.1.18.) This example is crucial for our further discussion:

$$\textbf{\textit{Mon}}(\textbf{\textit{Vect}}_\Bbbk) = \textbf{\textit{Alg}}_\Bbbk.$$

Also, with the canonical symmetry τ on $\textbf{\textit{Vect}}_\Bbbk$, we have

$$\textbf{\textit{cMon}}(\textbf{\textit{Vect}}_\Bbbk) = \textbf{\textit{cAlg}}_\Bbbk,$$

the category of commutative $\Bbbk$-algebras.

3.5.12 Graded algebras. The notion of commutativity depends on the symmetric structure. Consider the monoidal category $(\textbf{\textit{grVect}}_\Bbbk, \otimes, \Bbbk)$ of graded vector spaces (cf. 3.2.46). A monoid in $\textbf{\textit{grVect}}_\Bbbk$ is precisely a graded algebra (i.e. a vector space with a multiplication which respects the grading and whose neutral element is of degree 0).

We saw that there are two distinct interesting symmetries on $\textbf{\textit{grVect}}_\Bbbk$: the canonical one σ (induced from $\textbf{Vect}$) and the Koszul sign change κ. Let H be a graded-commutative algebra (i.e. $ab = (-1)^{pq}ba$, where $\deg(a) = p$ and $\deg(b) = q$). Then H is *not* commutative as monoid in $(\textbf{\textit{grVect}}_\Bbbk, \otimes, \Bbbk, \sigma)$, but it *is* commutative as monoid in $(\textbf{\textit{grVect}}_\Bbbk, \otimes, \Bbbk, \kappa)$. In fact, we can easily see that

$$\textbf{\textit{cMon}}(\textbf{\textit{grVect}}_\Bbbk, \kappa) = \textbf{\textit{gr-cAlg}}_\Bbbk$$

the category of graded-commutative algebras.

3.5.13 Trivial monoids. In any monoidal category $(\mathbf{V}, \Box, I)$ there is the trivial monoid I, where $\eta : I \to I$ and $\mu : I \Box I = I \to I$ are both the identity arrow of I.

3.5.14 Lemma. *Let* $(\mathbf{V}, \Box, I)$ *and* $(\mathbf{V}', \Box', I')$ *be monoidal categories, and let* $F : \mathbf{V} \to \mathbf{V}'$ *be a monoidal functor. If* (M, μ, η) *is a monoid in* $\mathbf{V}$ *then the image* $(MF, \mu F, \eta F)$ *is a monoid in* $\mathbf{V}'$.

Proof. That F is monoidal means in particular that $(M \Box M)F = MF \Box' MF$ and $IF = I'$. Using these identities we get maps

$$MF \Box' MF = (M \Box M)F \xrightarrow{\mu F} MF \qquad \text{and} \qquad I' = IF \xrightarrow{\eta F} MF.$$

Now we must verify that the monoid axioms are satisfied. Consider the diagram which expresses the associativity axiom for (M, μ, η) in $(\mathbf{V}, \Box, I)$ and apply F to it to get a diagram in $\mathbf{V}'$:

$$\begin{array}{ccc} (M\square M\square M)F & \longrightarrow & (M\square M)F \\ \downarrow & & \downarrow \\ (M\square M)F & \longrightarrow & MF \end{array}$$

The fact that F is monoidal means that this diagram can be identified with

$$\begin{array}{ccc} MF\square' MF\square' MF & \longrightarrow & MF\square' MF \\ \downarrow & & \downarrow \\ MF\square' MF & \longrightarrow & MF \end{array}$$

which is the diagram for the associativity axiom for $(MF, \mu F, \eta F)$. The unit axiom is verified similarly. □

Monoidal functors from the simplex category

3.5.15 The monoidal category generated by a monoid. Whenever we have a monoid (M, μ, η) in a monoidal category $(\mathbf{V}, \square, I)$ we can talk about the monoidal subcategory of $\mathbf{V}$ generated by M. Let us denote it $\langle M \rangle$. The objects of $\langle M \rangle$ are by definition $I, M, M\square M, M\square M\square M, \ldots$; for short let us put (for $n \in \mathbb{N}$)

$$M^n := \underbrace{M\square \ldots \square M}_{n \text{ factors}},$$

with $M^0 = I$. The arrows are all the arrows one can obtain by composition and monoidal operation on the structure maps of M,

$$\eta : I \to M \qquad \mathrm{id}_M : M \to M \qquad \mu : M\square M \to M.$$

If we draw M^n as n dots in a column then the pictures of these basic maps are

$\mu^{(0)}$ $\qquad$ $\mu^{(1)} = \mathrm{id}$ $\qquad$ $\mu^{(2)}$

and composition and monoidal operation then amount to serial and parallel connection, just as in our drawings of maps in Δ.

So all this looks a lot like Δ, and in a sense it is! We will soon make that statement precise.

3.5.16 Monoids in Δ. What are the monoids in $(\Delta, +, \mathbf{0})$? Well, to describe one we have to choose an object $\mathbf{n}$, a neutral element map $\eta : \mathbf{0} \to \mathbf{n}$, and a multiplication map $\mu : \mathbf{n} + \mathbf{n} \to \mathbf{n}$, and verify the axioms of 3.5.1. We already know from Example 3.5.13 that this is possible for $\mathbf{n} = \mathbf{0}$ (that gives the trivial monoid).

For $\mathbf{n} = \mathbf{1}$ things are fine as well. We know from 3.4.10 that there are unique maps $\mathbf{0} \to \mathbf{1} \leftarrow \mathbf{2} = \mathbf{1} + \mathbf{1}$, and the relations we found for these maps in 3.4.11 are then exactly the monoid axioms for $\mathbf{1}$, so there is a monoid $(\mathbf{1}, \mu, \eta)$. For convenience let us repeat the unit axiom:

=   =

Now we claim there are no further monoids in $(\Delta, +, \mathbf{0})$. Indeed let us try to construct one and see where the impossibility lies. For simplicity take $\mathbf{2}$ – the argument applies to any $n \geq 2$.

The two maps η and id for $\mathbf{2}$ certainly are

$\mathbf{0} \to \mathbf{2}$ $\mathbf{2} \to \mathbf{2}$

and

Now the 'multiplication' μ for $\mathbf{2}$ would be a map $\mathbf{2} + \mathbf{2} = \mathbf{4} \to \mathbf{2}$, and there is only one map that could satisfy the unit axioms:

=   =

and this map does not exist in Δ since crossing-over is not allowed! – it is not an order-preserving map. (In the bigger category Φ, the above map does exist, so $\mathbf{2}$ will be a monoid in Φ, as we shall see in 3.5.22.) In conclusion

$(\Delta, +, \mathbf{0})$ *contains a single nontrivial monoid, namely* $(\mathbf{1}, \mu, \eta)$.

Now what is the monoidal subcategory of $(\Delta, +, \mathbf{0})$ generated by $\mathbf{1}$? It is easy to see that that is $(\Delta, +, \mathbf{0})$ itself. Also, there are no relations other than exactly those required for having the monoid. Informally this is to say that

$(\Delta, +, \mathbf{0})$ *is the free monoidal category on a single monoid.*

3.5.17 Monoidal functors out of $(\Delta, +, \mathbf{0})$. A more precise meaning of the word 'free' is to have a certain universal property – in the present case this: *any monoid in any monoidal category* $\mathbf{V}$ *is the image of* $\mathbf{1}$ *under a unique monoidal functor* $\Delta \to \mathbf{V}$.

Just like the statement that $(\mathbb{N}, +, 0)$ is the free monoid generated by 1 means that any element in any monoid M is the image of 1 under a unique monoid homomorphism $\mathbb{N} \to M$, cf. 3.1.15. (Another example: the statement that $\Bbbk$ is the free vector space generated by 1 means that any vector in any vector space V is the image of 1 under a unique linear map $\Bbbk \to V$.)

In general, a monoidal functor is determined completely by its values on generators of the source category. In the case of Δ the generators are just the structure maps of the monoid **1**, so a monoidal functor out of Δ is determined completely by its value on the monoid **1**.

Now we claim that *given a monoidal category* $(\mathbf{V}, \Box, I)$ *there is a one-to-one correspondence between monoids in* $\mathbf{V}$ *and monoidal functors* $(\Delta, +, \mathbf{0}) \to (\mathbf{V}, \Box, I)$. In one direction the correspondence goes like this:

$$\boldsymbol{MonCat}(\Delta, \mathbf{V}) \leftrightarrow \boldsymbol{Mon}(\mathbf{V})$$
$$[\mathscr{M} : \Delta \to \mathbf{V}] \mapsto (\mathbf{1}\mathscr{M}, \mu\mathscr{M}, \eta\mathscr{M})$$

sending a monoidal functor $\mathscr{M}$ to its value on the monoid **1**. (The image of a monoid is again a monoid, cf. 3.5.14.) The functor 'points to' a specified monoid just like a monoid homomorphism $\mathbb{N} \to M$ points to an element of the monoid M.

In the other direction the correspondence goes like this: given a monoid M in $(\boldsymbol{V}, \Box, I)$, define a monoidal functor

$$\mathscr{M} : \Delta \longrightarrow \mathbf{V}$$
$$[\mathbf{0} \to \mathbf{1} \leftarrow \mathbf{2}] \longmapsto [I \to M \leftarrow M \Box M].$$

This makes sense because the relations that hold among the generators in Δ correspond exactly to the monoid axioms for M.

Since the sets involved in this correspondence are actually categories, we will strengthen the result a bit, to have it pronounce itself also on arrows:

3.5.18 Theorem. *There is a canonical equivalence of categories*

$$\boldsymbol{MonCat}(\Delta, \boldsymbol{V}) \simeq \boldsymbol{Mon}(\boldsymbol{V}),$$

where $\boldsymbol{Mon}(\boldsymbol{V})$ is the category of monoids in $\boldsymbol{V}$ and monoid homomorphisms, and $\boldsymbol{MonCat}(\Delta, \boldsymbol{V})$ is the category of monoidal functors from Δ to $\mathbf{V}$, and monoidal natural transformations.

3.5.19 Remark on equivalence versus isomorphism. The way we have set things up, with everything strict, the equivalence is in fact an *isomorphism* of categories. The reason for stating it as just an equivalence is to get a more robust statement, which remains true if the involved notions are weakened. If we relax the notion of monoidal categories to include nonstrict monoidal categories (cf. 3.2.15), then we must also relax the notion of monoidal functors (cf. 3.2.17) – otherwise $\boldsymbol{MonCat}(\Delta, \boldsymbol{V})$ might be empty! – and then the statement is an equivalence, not an isomorphism. This is a fundamental insight in category theory: that the good generalisation of 'bijection of sets' to the context of categories is 'equivalence of categories' (not 'isomorphism of categories').

Proof. Let us construct a functor $\Xi : \boldsymbol{MonCat}(\Delta, \boldsymbol{V}) \to \boldsymbol{Mon}(\boldsymbol{V})$. We have already explained what it does on objects: it takes a monoidal functor $F : \Delta \to \boldsymbol{V}$ to the monoid $\mathbf{1}F$.

Now we have to say what it does on arrows. Given two such monoidal functors, F and G, then a monoidal natural transformation $u : F \Rightarrow G$ is given by specifying its components $u_{\mathbf{n}} : \mathbf{n}F \to \mathbf{n}G$. That u is monoidal means (cf. 3.2.49) that the component $u_{\mathbf{1}} : \mathbf{1}F \to \mathbf{1}G$ determines all the others (for instance, $u_{\mathbf{2}} : \mathbf{2}F \to \mathbf{2}G$ must be the map $u_{\mathbf{1}} \Box u_{\mathbf{1}}$.) The naturality of u means that the maps $u_{\mathbf{n}}$ are compatible with all arrows in Δ. But the arrows in Δ are all generated by the structure maps for the monoid $\mathbf{1}$ in Δ, so the requirement amounts to saying that $u_{\mathbf{1}} : \mathbf{1}F \to \mathbf{1}G$ is a homomorphism of monoids in $\boldsymbol{V}$. So Ξ takes u to the monoid homomorphism $u_{\mathbf{1}}$.

Now we must check that the function Ξ we have defined is in fact a *functor*: we must check that it is compatible with composition of arrows, and also that it preserves identity. The composition of two natural transformations $F \overset{u}{\Rightarrow} G \overset{v}{\Rightarrow} H$ is just given by composing all the components $u_{\mathbf{n}}$ and $v_{\mathbf{n}}$. Since these in turn are determined completely by the first components $u_{\mathbf{1}}$ and $v_{\mathbf{1}}$, we see that composition of such natural transformations corresponds exactly to composition of monoid maps $\mathbf{1}F \overset{u_1}{\to} \mathbf{1}G \overset{v_1}{\to} \mathbf{1}H$, so Ξ is indeed a functor.

We leave it as an exercise to define the functor in the other direction, and to show that they are really inverses to each other. □

Algebras

In 3.5.11 we noted that

$$\boldsymbol{Mon}(\boldsymbol{Vect}_{\Bbbk}) = \boldsymbol{Alg}_{\Bbbk},$$

a monoid in $(\boldsymbol{Vect}_{\Bbbk}, \otimes, \Bbbk)$ is nothing but an algebra. Combining this with Theorem 3.5.18, we find

3.5.20 Corollary. *There is a canonical equivalence of categories*

$$\boldsymbol{MonCat}(\Delta, \boldsymbol{Vect}_{\Bbbk}) \simeq \boldsymbol{Mon}(\boldsymbol{Vect}_{\Bbbk}) = \boldsymbol{Alg}_{\Bbbk}.$$

If we recall that Δ could be characterised as a certain category of noncrossing strands, just like ***2Cob*** is described in terms of certain tubes, we see that this result is analogous to 3.3.2.

A more precise analogy arises when we pass to the symmetric case:

Symmetric monoidal functors on Φ

Copying over the analysis we did in 3.5.16 it is easy to find all the monoids in $(\Phi, +, \mathbf{0})$, but the conclusion turns out to be radically different. The reason is that $+$ is the coproduct in Φ, cf. 3.4.19.

It is just as easy to prove this general result:

3.5.21 Lemma. *In a monoidal category given by coproduct, every object carries a unique monoid structure, and this monoid structure is commutative.*

Proof. Let $\mathbf{V}$ be a category with coproduct $\Box$ and initial object I. Then $(\mathbf{V}, \Box, I)$ is a monoidal category with a unique symmetry τ (cf. 3.2.42). Let M be an object in $\mathbf{V}$. Since I is initial object there is a unique arrow $\eta : I \to M$. The question is whether there is an arrow $\mu : M\Box M \to M$ which has η as neutral from both sides, and which is associative. Now by the universal property of the coproduct, giving an arrow $\mu : M\Box M \to M$ is equivalent to giving two arrows $\mu_1, \mu_2 : M \to M$. But the left neutral requirement says that $I\Box M \to M\Box M \to M$ is the identity, so μ_2 must be the identity, and similarly the right neutral axiom implies that μ_1 is the identity, so the arrow $\mu : M\Box M \to M$ is the identity on each 'factor'. It is easy to see that μ is associative and commutative. $\Box$

3.5.22 Monoids in Φ. By the lemma, *every object* $\mathbf{n}$ *in* Φ *admits a unique monoid structure and this structure is commutative.* Comparing with the situation in $(\Delta, +, 0)$, the trivial monoid $\mathbf{0}$ looks of course the same; the pictures for $\mathbf{1}$ are also very much as in Δ, but now it is commutative:

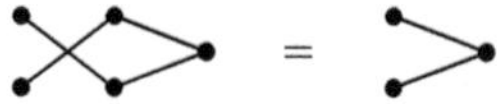

The monoid $\mathbf{2}$ has no analogue in Δ: the multiplication map is

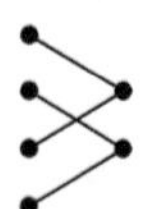

and the commutativity of this monoid is expressed by the drawing

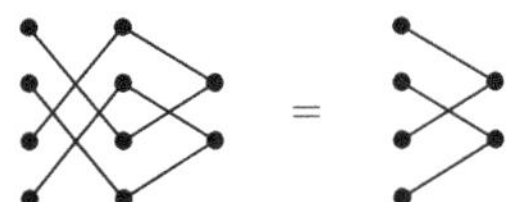

3.5.23 The symmetric monoidal category generated by a commutative monoid S is the category whose objects are all the powers of S, and whose arrows are all the maps you can get from the structure maps of S together with the twist maps of those objects.

Each of the monoids in Φ generates a symmetric monoidal category. The symmetric monoidal category generated by $\mathbf{1}$ is the whole of $(\Phi, +, \mathbf{0})$. The symmetric monoidal category generated by $\mathbf{2}$ contains all the objects $\mathbf{2n}$ and all the arrows between such objects which consist of 'closely parallel strands'. (Exercise: explain what this is supposed to mean!) For example it contains the twist map $\mathbf{2} + \mathbf{2} \to \mathbf{2} + \mathbf{2}$:

but not

(The only arrow $\mathbf{2} \to \mathbf{2}$ is the identity.)

We see that each object generates a monoidal subcategory in $(\Phi, +, \mathbf{0})$ isomorphic to $(\Phi, +, \mathbf{0})$. This is the categorification of the fact that each element n in the monoid $(\mathbb{N}, +, 0)$ generates a monoid $\langle n \rangle$ isomorphic to $(\mathbb{N}, +, 0)$. (Note that when a symmetric monoidal category contains a monoid then it automatically contains many: all the tensor powers are also monoids. This is a variation of 3.5.5.)

3.5.24 Theorem. $(\Phi, +, \mathbf{0})$ *is the free symmetric monoidal category on a commutative monoid, in the sense that for each symmetric monoidal category* $\mathbf{V} = (\mathbf{V}, \Box, I, \tau)$ *there is a canonical equivalence of categories*

$$\textbf{\textit{SymMonCat}}(\Phi, \mathbf{V}) \simeq \textbf{\textit{cMon}}(\mathbf{V}).$$

This is just a variation of 3.5.18 – the remark given after that theorem applies equally here in the symmetric context.

The proof is also basically the same. Given a symmetric monoidal functor $\Phi \to \mathbf{V}$ then the image of $\mathbf{1}$ is a commutative monoid in $\mathbf{V}$ (this is an easy symmetric analogue of Lemma 3.5.14). On the other hand, its values on objects are determined completely by its value on $\mathbf{1}$, since it is a monoidal functor. Concerning arrows: given a monoidal natural transformation between two such functors $u : \mathscr{M} \to \mathscr{N}$, then $u_1 : 1\mathscr{M} \to 1\mathscr{N}$ is a homomorphism of monoids, (this follows from naturality of u with respect to the generators of Φ). Conversely, a homomorphism of commutative monoids determines uniquely a natural transformation between the corresponding functors: the point is that the relations in Φ correspond precisely to the axioms for a commutative monoid.

In particular, with $\mathbf{V} = (\textit{\textbf{Vect}}_{\Bbbk}, \otimes, \Bbbk, \sigma)$ we have

3.5.25 Corollary. *There is a canonical equivalence of categories*

$$\textit{\textbf{Repr}}_{\Bbbk}(\Phi) = \textit{\textbf{SymMonCat}}(\Phi, \textit{\textbf{Vect}}_{\Bbbk}) \simeq \textbf{cMon}(\textit{\textbf{Vect}}_{\Bbbk}) = \textbf{cAlg}_{\Bbbk}.$$

3.5.26 Example. In the symmetric monoidal category $(\textit{\textbf{2Cob}}, \coprod, \varnothing, T)$, the circle is a commutative monoid. It is the image of the symmetric monoidal functor $(\Phi, +, 0, \tau) \to (\textit{\textbf{2Cob}}, \coprod, \varnothing, T)$ defined in the obvious way: $\mathbf{1}$ is sent to the circle (which we have already been denoting by $\mathbf{1}$ for a while!) – and the remaining data are completely determined by the requirement that the functor be monoidal and symmetric.

Consider now the forgetful functor $\textbf{cFA}_{\Bbbk} \to \textbf{cAlg}_{\Bbbk}$ (forgetting the Frobenius structure). By Theorem 3.3.2, a given commutative Frobenius algebra (A, ε) corresponds to a symmetric monoidal functor $\textit{\textbf{2Cob}} \to \textit{\textbf{Vect}}_{\Bbbk}$, and we can compose with the symmetric monoidal functor $\Phi \to \textit{\textbf{2Cob}}$ to get one $\Phi \to \textit{\textbf{Vect}}_{\Bbbk}$. By Theorem 3.5.24, this corresponds to a commutative monoid in $\textit{\textbf{Vect}}_{\Bbbk}$ (i.e. a commutative algebra), and it is easy to verify that this algebra is just A.

3.5.27 Example. In the same vein, a commutative monoid M in some symmetric monoidal category $(\mathbf{V}, \square, I, \tau)$ is given by a symmetric monoidal functor $\mathscr{M} : \Phi \to \mathbf{V}$. Since M is a monoid, by Theorem 3.5.18 it corresponds to a monoidal functor $\Delta \to \mathbf{V}$. This functor is just the composite

$$\Delta \longrightarrow \Phi \xrightarrow{\mathscr{M}} \mathbf{V}.$$

Exercises

1. Let $(\mathbf{V}, \square, I)$ be a monoidal category, and let $\phi : M \to M'$ be a homomorphism of monoids in $\mathbf{V}$. Show that if ϕ is invertible as an arrow in $\mathbf{V}$, then it is also invertible as a monoid homomorphism (and conversely). (This generalises Exercise 2 on page 148.) (Hint: use Exercise 2 on page 168.)

2. *The Eckmann–Hilton argument.* Since the monoids in ***Set*** form a monoidal category $\textbf{\textit{Mon}} = \textbf{\textit{Mon}}(\textbf{\textit{Set}})$ under cartesian product (cf. 3.2.24, 3.5.5), it makes sense to speak about monoids in here.

   (i) Show that a monoid in $(\textbf{\textit{Mon}}, \times, 1)$ is the same as a set M with two compatible monoid structures $*_{\mathrm{h}}$ and $*_{\mathrm{v}}$ (referred to as *horizontal* and *vertical*). The compatibility requirements are these: the unit should be the same for $*_{\mathrm{h}}$ and $*_{\mathrm{v}}$, and furthermore, for all $a, b, c, d \in M$:

   $$(a *_{\mathrm{h}} b) *_{\mathrm{v}} (c *_{\mathrm{h}} d) = (a *_{\mathrm{v}} c) *_{\mathrm{h}} (b *_{\mathrm{v}} d).$$

   This is called the *interchange law* or *Godement relation*. Remark: the interchange law actually implies that the two units coincide, and in fact it implies associativity for each of the multiplication laws too! – write down the proof of these two statements.

   Here is an illustration you can try to interpret:

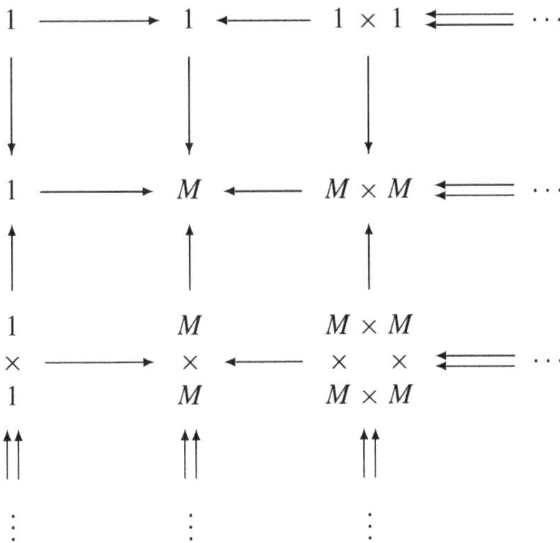

   and here is a drawing meant to illustrate the interchange law:

   $$\boxed{\begin{array}{c} a\ b \\ \hline c\ d \end{array}} = \boxed{\begin{array}{c|c} a & b \\ c & d \end{array}}$$

   (ii) Show that these two monoid structures $*_{\mathrm{h}}$ and $*_{\mathrm{v}}$ coincide, and that this monoid is commutative. Here is an illustration for your proof:

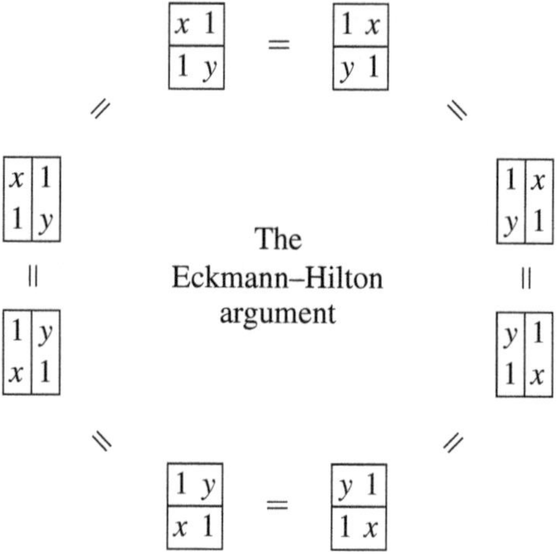

(iii) Go through all the arguments above and explain where the canonical symmetry on (***Set***, $\times$, 1) was used, and where the twist map was hidden.

3.6 Frobenius structures

Comonoids and coalgebras

3.6.1 Comonoids. A *comonoid* in a monoidal category $(\boldsymbol{V}, \square, I)$ is an object M equipped with a *comultiplication* $\delta : M \to M \square M$ and a *counit* $\varepsilon : M \to I$,

δ ε

satisfying the axioms dual to the monoid axioms of 3.5.1:

A *comonoid homomorphism* between two comonoids in $\boldsymbol{V}$ is one which preserves the comonoid structure. There is a category ***Comon***$(\boldsymbol{V})$ of comonoids and comonoid homomorphisms in $\boldsymbol{V}$.

Since the arrows defining a comonoid are just reversed compared to the arrows defining a monoid, to give a comonoid in $\boldsymbol{V}$ amounts to giving a monoid in $\boldsymbol{V}^{\mathrm{op}}$. To give a comonoid homomorphism between comonoids in $\boldsymbol{V}$ is to give a monoid homomorphism in the other direction between the corresponding monoids in $\boldsymbol{V}^{\mathrm{op}}$. So we have ***Comon***$(\boldsymbol{V}) \simeq$ ***Mon***$(\boldsymbol{V}^{\mathrm{op}})^{\mathrm{op}}$.

If $\boldsymbol{V}$ is a *symmetric* monoidal category then $\boldsymbol{Comon}(\boldsymbol{V})$ is a monoidal category. This follows from the arguments of 3.5.5.

3.6.2 Cocommutative comonoids. A comonoid in a symmetric monoidal category $(\boldsymbol{V}, \square, I, \tau)$ is called *cocommutative* if it satisfies the axiom dual to the commutativity axiom 3.5.6:

The category of cocommutative comonoids in $\boldsymbol{V}$ is denoted $\boldsymbol{cComon}(\boldsymbol{V})$.

3.6.3 Lemma. *In a monoidal category whose monoidal structure is given by product, each object admits a unique comonoid structure (which is cocommutative).*

This statement is just dual to 3.5.21.

3.6.4 Example: comonoids in *Set*. In the monoidal category $(\boldsymbol{Set}, \times, 1)$, every set S has a unique comonoid structure, namely the one whose comultiplication map is the diagonal map $S \to S \times S$. (The counit map is the unique $S \to 1$.)

3.6.5 Coalgebras. A comonoid in $(\boldsymbol{Vect}_{\Bbbk}, \otimes, \Bbbk)$ is precisely a coalgebra (see 2.3.1):

$$\boldsymbol{Comon}(\boldsymbol{Vect}_{\Bbbk}) = \boldsymbol{Coalg}_{\Bbbk}.$$

(The proof is completely analogous (in fact dual) to the proof of 3.5.11.)

Also,

$$\boldsymbol{cComon}(\boldsymbol{Vect}_{\Bbbk}) = \boldsymbol{cCoalg}_{\Bbbk}.$$

3.6.6 The universal comonoid. The monoidal category Δ^{op} contains a unique nontrivial comonoid, namely $\mathbf{1}$. (This statement is just dual to 3.5.16.) In fact, Δ^{op} is the free monoidal category on a comonoid, in the sense that having a comonoid in a monoidal category $\boldsymbol{V}$ is just like having a monoidal functor $\Delta^{\mathrm{op}} \to \boldsymbol{V}$. Similarly, Φ^{op} is the free symmetric monoidal category containing a cocommutative comonoid. We can write

$$\boldsymbol{MonCat}(\Delta^{\mathrm{op}}, \boldsymbol{V}) \simeq \boldsymbol{Comon}(\boldsymbol{V})$$
$$\boldsymbol{SymMonCat}(\Phi^{\mathrm{op}}, \boldsymbol{V}) \simeq \boldsymbol{cComon}(\boldsymbol{V}).$$

In particular,

$$\textbf{\textit{MonCat}}(\Delta^{\text{op}}, \textbf{\textit{Vect}}_{\Bbbk}) \simeq \textbf{\textit{Coalg}}_{\Bbbk}$$
$$\textbf{\textit{Repr}}_{\Bbbk}(\Phi^{\text{op}}) = \textbf{\textit{SymMonCat}}(\Phi^{\text{op}}, \textbf{\textit{Vect}}_{\Bbbk}) \simeq \textbf{\textit{cCoalg}}_{\Bbbk}.$$

3.6.7 Digression on bimonoids and bialgebras. A *bimonoid* in a symmetric monoidal category $(V, \square, I, \tau)$ is an object B which is simultaneously a monoid and a comonoid, and such that the following compatibility conditions hold:

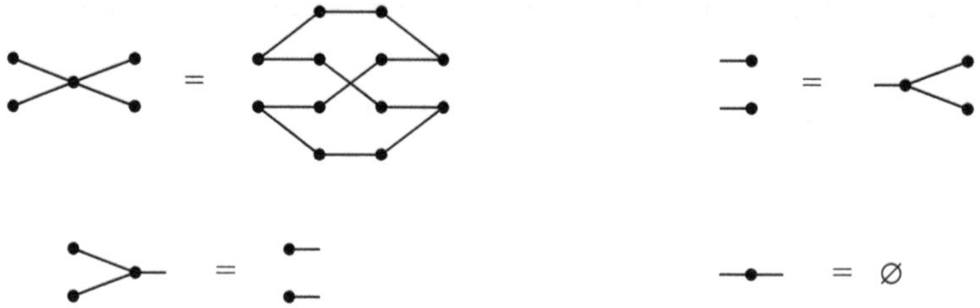

This definition is made in order for bimonoids in $(\textbf{\textit{Vect}}_{\Bbbk}, \otimes, \Bbbk, \sigma)$ to be exactly bialgebras (cf. 2.4.9) – this is easy to see comparing with the figures drawn there. For algebras we observed that these axioms amount to saying that the comultiplication and counit maps are algebra homomorphisms (or equivalently, that multiplication and unit are coalgebra homomorphisms). Similarly in this general setting, a bimonoid is a monoid equipped with a comultiplication and a counit which are monoid homomorphisms (or equivalently, a comonoid equipped with a multiplication and a unit which are comonoid homomorphisms). In other words

$$\begin{aligned} \textbf{\textit{Bimon}}(V) &\simeq \textbf{\textit{Comon}}(\textbf{\textit{Mon}}(V)) \\ &\simeq \textbf{\textit{Mon}}(\textbf{\textit{Comon}}(V)). \end{aligned}$$

Note that since V is a symmetric monoidal category, $\textbf{\textit{Mon}}(V)$ is a monoidal category, cf. 3.5.5. A similar observation holds of course for $\textbf{\textit{Comon}}(V)$ by duality, so the two statements make sense.

Frobenius objects, Frobenius algebras, and 2-dimensional cobordisms

The bimonoids of the previous paragraph were only mentioned as a contrast: what we really want is another compatibility between multiplication and comultiplication, namely the Frobenius condition. So we introduce the notion of Frobenius object in a general monoidal category, in such a way that Frobenius objects in ***Vect*** be precisely Frobenius algebras:

3.6.8 Frobenius objects. A *Frobenius object* in a monoidal category $(\boldsymbol{V}, \Box, I)$ is an object A equipped with four maps:

$\eta : I \to A$ $\mu : A \Box A \to A$ $\delta : A \to A \Box A$ $\varepsilon : A \to I$

satisfying the unit and counit axioms:

as well as the *Frobenius relation*:

3.6.9 Frobenius algebras. A Frobenius object in $(\boldsymbol{Vect}_{\Bbbk}, \otimes, \Bbbk)$ is precisely a Frobenius algebra. This is the content of 2.3.24. Note that a priori we did not require vector spaces to be of finite dimension, but we have shown in 2.1.12 that the Frobenius condition on a vector space implies finite dimension.

3.6.10 Lemma. *The multiplication μ of a Frobenius object A is associative, and the comultiplication δ is coassociative. In other words, A is at the same time a monoid and a comonoid.*

Proof. This was proved for Frobenius algebras in 2.3.24. Since the proof was completely graphical, it is valid for Frobenius objects in any monoidal category. □

3.6.11 Frobenius homomorphisms. A *Frobenius homomorphism* between two Frobenius objects in $\boldsymbol{V}$ is a map that preserves all the structure, that is, a map which is at the same time a monoid homomorphism and a comonoid homomorphism. There is a category $\boldsymbol{Frob}(\boldsymbol{V})$ of Frobenius objects and Frobenius homomorphisms in $\boldsymbol{V}$. So 3.6.9 can more precisely be stated as

$$\boldsymbol{Frob}(\boldsymbol{Vect}_{\Bbbk}) = \boldsymbol{FA}_{\Bbbk}.$$

3.6.12 Question. We saw in 2.4.5 that a homomorphism of Frobenius algebras is always invertible. Is the same true for a Frobenius homomorphism in a general monoidal category? (This seems difficult because the argument given there used kernels and dimensions and other vector space specific features...)

3.6.13 Commutative Frobenius objects. It is obvious what commutativity should mean for a Frobenius object – provided it lives in a symmetric monoidal category! A Frobenius object in a symmetric monoidal category $(\boldsymbol{V}, \Box, I, \tau)$

is said to be *commutative* if it is commutative as a monoid. So the defining condition is

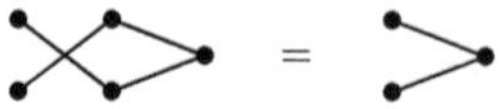

(Clearly there is a category **cFrob**$(\boldsymbol{V})$ of commutative Frobenius objects and Frobenius homomorphisms in $\boldsymbol{V}$.) Hence the commutative Frobenius objects in $(\mathbf{Vect}_{\Bbbk}, \otimes, \Bbbk, \sigma)$ are precisely the commutative Frobenius algebras.

Similarly we can define what it means to be *cocommutative*. The condition is:

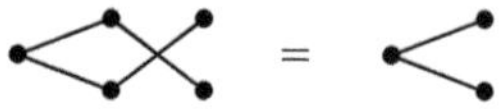

3.6.14 Lemma. *A Frobenius object in a symmetric monoidal category is cocommutative if and only if it is commutative.*

Proof. We saw in 2.3.29 that this result is true when the symmetric monoidal category is $(\mathbf{Vect}_{\Bbbk}, \otimes, \Bbbk, \sigma)$. Since the proof was graphical it carries over to the case of an arbitrary symmetric monoidal category. □

3.6.15 Example. In the symmetric monoidal category $(\mathbf{2Cob}, \coprod, \varnothing, T)$, every object has a canonical structure of a commutative Frobenius object. In particular, **1** is a commutative Frobenius object – this is more or less the content of the description in terms of generators and relations given in Chapter 1.

(More generally, the $(n-1)$-sphere is a commutative Frobenius object in $\boldsymbol{n}\mathbf{Cob}$, and so is any disjoint union of $(n-1)$-spheres.)

Now just as we found that Δ is the free monoidal category on a monoid, and Φ is the free symmetric monoidal category on a commutative monoid, it is now natural to look for similar universal Frobenius structures: what is the free monoidal category containing a Frobenius object? – and what about the free symmetric monoidal category on a commutative Frobenius object?

3.6.16 The free monoidal category on a Frobenius object. The obvious way of describing this universal Frobenius structure – the free monoidal category on a Frobenius object – is in terms of generators and relations: simply take as little as possible. Let $(\boldsymbol{X}, \square, \mathbf{0})$ be the monoidal category whose objects are the powers $\mathbf{n} := \mathbf{1}\square \ldots \square\mathbf{1}$ of a given object **1** which we picture as a dot, and whose set of arrows is generated by these four:

and subject to the three relations (unit, counit, and Frobenius):

Then clearly $\mathbf{1}$ is a Frobenius object in $\boldsymbol{X}$. With this definition, the proof of 3.3.2 carries over directly to establish

3.6.17 Theorem. *For every monoidal category* $(\mathbf{V}, \square, I)$ *there is a canonical equivalence of categories*

$$\boldsymbol{MonCat}(\boldsymbol{X}, \boldsymbol{V}) \simeq \boldsymbol{Frob}(\boldsymbol{V}).$$

3.6.18 The category of 2-cobordisms. Next we apply the same arguments to the symmetric case. Define a symmetric monoidal category by taking its generators to be the four maps above, together with the twist map, and subject to unit, counit, and Frobenius relations, imposing furthermore all the relations involving the twist map (i.e. the axioms for being a symmetry structure, cf. 3.2.34; see also 3.4.26), and impose finally the commutativity relation (cf. 3.6.13). (Then $\mathbf{1}$ is clearly a commutative Frobenius object.)

But this category was amply described in Chapter 1: it is precisely (the skeleton of) ***2Cob***! The only difference is notational: in Chapter 1 we used thicker tube-like symbols instead of the skinny pictures used in this chapter. (Also, in Chapter 1 we listed more relations, for example the associativity and coassociativity relations, but we know from 3.6.10 that these relations are redundant once we have the Frobenius relation.) Now the arguments of the proof of Theorem 3.3.2 carry over word for word to prove more generally:

3.6.19 Theorem. *The skeletal cobordism category* $(\mathbf{2Cob}, \coprod, \emptyset, T)$ *is the free symmetric monoidal category containing a (co)commutative Frobenius object. In other words, given a (co)commutative Frobenius object A in a symmetric monoidal category* $(\mathbf{V}, \square, \boldsymbol{I}, \tau)$ *then there is a unique symmetric monoidal functor* $\mathbf{2Cob} \to \mathbf{V}$ *such that* $\mathbf{1} \mapsto A$. *This gives a canonical equivalence of categories*

$$\boldsymbol{SymMonCat}(\boldsymbol{2Cob}, \boldsymbol{V}) \simeq \boldsymbol{cFrob}(\boldsymbol{V}).$$

This is the happy ending of our movie – the well deserved and definitive reunion of the commutative Frobenius with the princess ***2Cob***. The remaining few paragraphs just tell the curious story about what happened to the hero's best friend, the noncommutative Frobenius, and how he also got married. (This will only be a brief outline – there is actually some work to do here. . .)

3.6.20 Towards a geometrical description of *X*. What makes Theorem 3.6.19 interesting compared to 3.6.17 is that ***2Cob*** was originally defined in purely geometrical terms. It is natural to ask for a similar geometric description of $\boldsymbol{X}$ – to see whether we can give flesh and bones to the abstract description given in terms of generators and relations.

The first approach to this problem is simply to take literally the drawings of the generators, and see what sort of geometric figures we can build out of them. The objects are columns of dots – lining up the dots in columns reflects the monoidal operation $\square$. We allow the empty column (the object $\mathbf{0}$), so there is an object for each natural number.

The arrows should be certain collections of strands between these dots, and since the two main operations (μ and δ) are not (co)commutative, we impose the rule that the strands are not allowed to cross over each other.

So far the description is just copied over from that of Δ. Just as for Δ, to be pedantic about the interpretation of the no-crossing rule we should require the whole picture to take place inside a rectangle, with the input dots on the left-hand side of the rectangle and the output column on the right.

From here on, the description diverges from that of Δ and things become more complicated. Since we dispose of the comultiplication δ

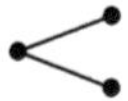

it is no longer true that exactly one strand emanates from each input dot – in fact, by composing several copies of δ we can have any number of strands emanating from a single input dot. Another way of saying this is that the strands are not only allowed to merge, they are also allowed to split. A notable consequence of this observation is that it will not be possible to interpret the strands as functions of any kind, as we did for Δ. . . We can even have an input dot from which no strands come out – this is the interpretation of the counit ε. We may choose to draw it instead as a little line sticking out and going nowhere, in accordance with our graphical convention of this chapter; in that case we should rather say that there is a always at least one strand coming out of each input dot, but that it does not have to go anywhere. . .

Next, consider the composite $\mu\varepsilon : \mathbf{2} \to \mathbf{0}$

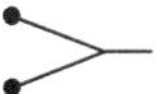

which in terms of Frobenius algebras is the pairing β. We are only interested in the combinatorics of these strands, not their particular shape, so we will rather picture this map as

(In a moment we will be more precise about the equivalence relation we are tacitly imposing...) So the strands are allowed to double back. Dually the composite $\eta\delta : \mathbf{0} \to \mathbf{2}$ provides the possibility of a strand like this

Next consider the composite $\eta\varepsilon : \mathbf{0} \to \mathbf{0}$. This is *not* the identity map on $\mathbf{0}$. For this reason it seems safer to write the small lines sticking in and out – then the drawing of this strand becomes

$$— \tag{3.6.21}$$

So it is a strand coming from nowhere and going nowhere, but it is nevertheless there, and it is not equivalent to the drawing where it is not there. (Perhaps you will enjoy imaging this as some exotic particle which appears and disappears again...)

More complicated examples of this kind can be built up by inserting composites $\delta\mu$ in the middle. We get this example of an arrow from the empty column to itself

Or even this:

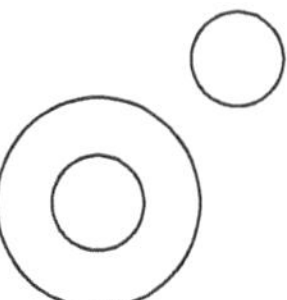

(Here we see a phenomenon which does not have any analogue in the symmetric case: nesting. Two nested circles do not represent the same arrow as two non-nested ones...)

So in conclusion, an arrow is nearly about any possible drawing you can imagine in the space between the two columns of dots!

3.6.22 Combinatorics of these drawings. We need to be precise about when we consider two such drawings to define the same arrow. For Δ that was easy:

we just invoked the interpretation of the strands as functions, the shape of the strands did not matter, only which dots they connected. In the present context, we cannot argue in terms of functions, but it is nevertheless clear that we are not interested in the precise shape of a strand, but only in its combinatorics. The combinatorics in this case is not only a question of which input dots are connected to which output dots, but concerns also what happens in between, namely the incidence relations between edges and inner vertices. Here *inner vertex* means any point where three or more curve segments meet, and *edge* means the curve segments, so for example in this drawing there are two inner vertices and five edges:

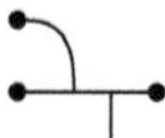

In other words, we are interested (at most) in the graph represented by the drawing. (The no-crossings axiom then says that we are dealing with planar graphs.)

This observation is somewhat unrelated to the relations that characterise $\boldsymbol{X}$; it is merely a question of killing the arbitrariness of the graphical representation of a structure which from the beginning was combinatorial. Now we come to

3.6.23 Interpretation of the relations. The unit and counit relations tell us that whenever we have an edge which connects to only one vertex, then we can contract it back to the vertex (whose valence is then decremented by 1, and it may even cease to be a vertex). For example in the previous drawing, the little vertical line segment can be removed, whereafter there is only one vertex left, and three edges. We already used this operation when drawing $\mu\varepsilon$

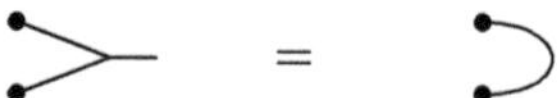

Note that a dead-end edge like this can be contracted (deleted) only if it is connected to a vertex (since this is the situation the unit and counit relations pronounce themselves about). The case of an edge without any vertices (number 3.6.21 above) is not covered by this rule and cannot be deleted.

Next we should notice that the Frobenius relation together with the unit and counit relations imply the snake relation

This was actually used when we stated that the shape of the edges is immaterial...

Finally, with the Frobenius relation we can really do some drastic operations on these graphs: we can move around with the vertices along the edges, and even move them past each other. For example in this animation

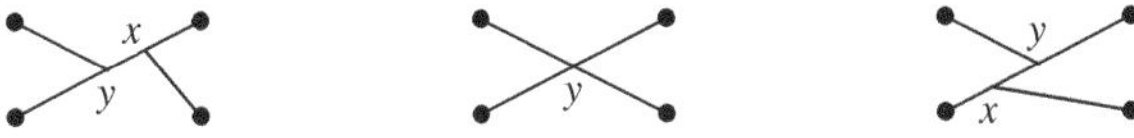

we see the vertex x move downwards and leftwards along the edge, coincide with the vertex y, and separate from it again, continuing past it on the other side.

Thus in particular we find this relation

Another way of describing the operations expressed by the Frobenius relation is to say that we can contract any edge between two distinct (inner) vertices. The two vertices then become one vertex (with total valence $m + n - 2$, if the original vertices had valences m and n). As a consequence, any connected graph with more than one inner vertex can be transformed into a graph with only one inner vertex. For example

3.6.24 The category of isotopy classes of planar graphs. The precise description of when we consider two graphs equivalent is given by the notion of isotopy of planar graphs. We say two planar graphs G_0, G_1 are *isotopic* if one can be deformed continuously into the other – without leaving the plane, i.e. there is a continuously varying family of graphs parametrised by the unit interval such that the graph over 0 is G_0 and the graph over 1 is G_1, and such that all the intermediate graphs are also planar. So the rules are: no breaking, no joining, no crossing over.

Now, the relations defining a Frobenius object express exactly isotopy invariance of certain modifications of planar graphs. So we can say that in the geometric description of $\boldsymbol{X}$ the arrows are *isotopy classes of planar graphs*.

The isotopy classes of connected graphs are classified by the number of input dots and output dots and the genus of the graphs (i.e. the number of loops). For nonconnected graphs, some invariant which takes care of nesting is required. . .

3.6.25 Thickening: 'planar cobordisms'. If we thicken all the pictures a little bit we get tubes. That leads to the plumber's description of the category $\boldsymbol{X}$. It is more or less like ***2Cob***, but we must prohibit the twist cobordism. So we are talking about a notion of 'planar cobordisms': systems of tubes which lie flat on the floor...

3.6.26 Comparison. With the above heuristic discussions, the reader is hopefully convinced that the geometrical description of ***X*** agrees with the one given in terms of generators and relations. Establishing a formal proof involves first of all more precise formulation of some of the concepts introduced above, and then imitation of the analysis we carried out for the category ***2Cob*** in Section 1.4 – chopping up any such graph in pieces isotopic to the four generators, defining a 'normal form', and so on. Some new concepts are needed: e.g. some sort of Morse theory for graphs, and a way of treating nonconnected graphs... This is left as a challenge to the reader.

3.6.27 A project... We have seen that the commutative Frobenius structure is topologically something not-embedded-anywhere, while the corresponding noncommutative structure is something embedded in the plane.

There is an interesting intermediate possibility, namely embedding in $\mathbb{R}^3$! We would still require the objects to be dots arranged along line segments in $\mathbb{R}^3$, and the strands should be in a specified box. But now they could cross over and under, and that would not be the same thing – we would have braids! So this construction should lead to something like the universal braided Frobenius structure, and there would arise some sort of notion of braided Frobenius algebra. In terms of generators and relations, in addition to the four basic maps η, μ, δ, and ε (cf. 3.6.8), there should be a *braid map* corresponding to crossing over (and its inverse: crossing under), satisfying the axioms for being a braiding in a braided monoidal category. The definition of braiding was given on page 169, and for all further information we refer to Kassel's book [29]. To work this out is beyond the scope of these notes, but the reader is encouraged to have a look at this problem on her own...

Appendix: vocabulary from category theory

A.1 Categories

A.1.1 Categories. A category $\boldsymbol{C}$ consists of

- a class of *objects* $\boldsymbol{C}_0$,
- for every pair of objects X, Y, a set $\boldsymbol{C}(X, Y)$ of *arrows* from X to Y,
- for every triple of objects X, Y, Z, an *associative composition law*

$$\boldsymbol{C}(X, Y) \times \boldsymbol{C}(Y, Z) \to \boldsymbol{C}(X, Z),$$

- and for every object X, a specified *identity arrow* $\mathrm{Id}_X \in \boldsymbol{C}(X, X)$ which acts as neutral element for the composition (from both sides).

We write $f : X \to Y$ for an element in $\boldsymbol{C}(X, Y)$. (The set $\boldsymbol{C}(X, Y)$ is often denoted $\mathrm{Hom}_{\boldsymbol{C}}(X, Y)$.) Given two arrows $f : X \to Y$ and $g : Y \to Z$ we write fg for their composite $X \to Y \to Z$. The associativity of the composition law means that given three arrows $W \xrightarrow{e} X \xrightarrow{f} Y \xrightarrow{g} Z$ then

$$(ef)g = e(fg).$$

Saying that $\mathrm{Id}_X : X \to X$ is neutral for the composition law from both sides means that for every arrow $f : X \to Y$ we have $\mathrm{Id}_X\, f = f$ and for every arrow $e : W \to X$ we have $e\, \mathrm{Id}_X = e$.

As short-hand notation in the next couple of diagrams and figures, let us write $\boldsymbol{C}(X, Y, Z) := \boldsymbol{C}(X, Y) \times \boldsymbol{C}(Y, Z)$ – this notation is not standard and is not used elsewhere in the text.

The associativity and neutral arrow axioms can be expressed by the following commutative diagrams:

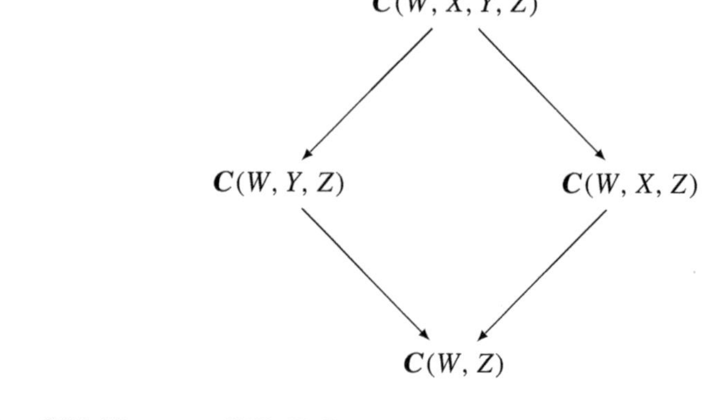

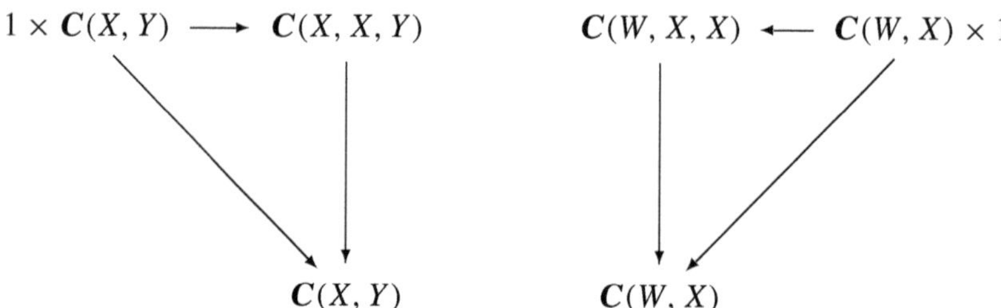

where 1 denotes a set with only one element, and $1 \to \boldsymbol{C}(X, X)$ is the map that sends this element to Id_X.

If you have already read some portions of the main text, perhaps you can have fun interpreting the following drawing – otherwise you should just ignore it:

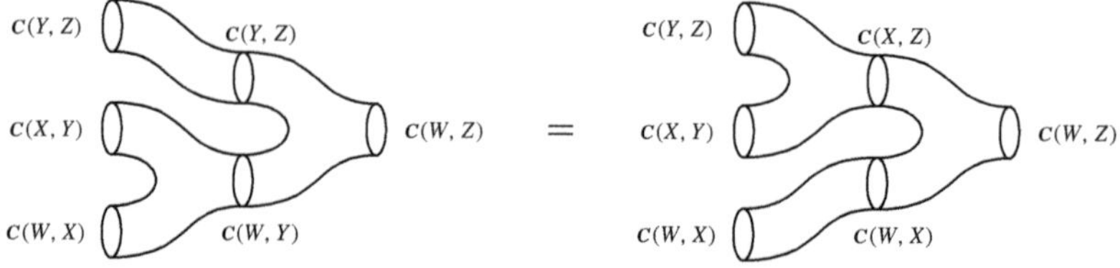

A.1.2 Large examples. The category ***Set*** whose objects are all sets and whose arrows are the set mappings. Composition is the obvious composition of mappings. Or the category ***FinSet*** of finite sets and set mappings between finite sets. This is a *subcategory* in ***Set***.

The category $\boldsymbol{Vect}_{\Bbbk}$ of vector spaces over a field $\Bbbk$, and $\Bbbk$-linear maps.

The category ***Top*** of topological spaces and continuous maps.

The category ***Grp*** of groups and group homomorphisms.

The category ***Ring*** of rings and ring homomorphisms.

The category ***FinOrd*** of finite ordered sets and order-preserving maps (see 3.4.1).

All the above examples are of this type: the objects are sets with some additional structure, and the arrows are maps which preserve this structure – such categories are called concrete. Here is an example which does not fit into this picture:

A.1.3 Groups as categories. Let G be a group. Define a category $\boldsymbol{G}$ with a single object x, by setting $\boldsymbol{G}(x, x) := G$. So the arrows are the elements of the group G, and composition is given by the composition law of the group. The neutral element of the group gives the identity arrow of the unique object x.

A.1.4 Smaller examples. Let **1** denote the category with just one object, and just one arrow (the identity arrow of that object).

Consider the category **2** with two objects, and a single arrow in addition to the two required identity arrows. It looks like this: $\bullet \to \bullet$. (The identity arrows have been omitted in the drawing.)

Consider the category **3** with three objects, and arrows

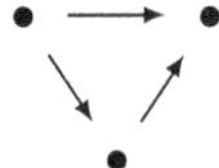

More generally there is a category **n** whose arrows are the edges of an oriented n-simplex – see 3.4.5.

Consider the category $\bullet \leftarrow \bullet \rightarrow \bullet$. Let us call this category $\boldsymbol{P}$ since we will use it again in A.3.3. There are no nontrivial compositions in this category.

Exercises

1. Show that identity arrows are unique for each object, i.e. if $a : X \to X$ and $b : X \to X$ both satisfy the identity arrow axiom then they are equal.
2. An arrow $i : X \to Y$ in a category $\boldsymbol{C}$ is called an *monomorphism* if it is cancellable from the right: this means that given a diagram

$$A \overset{a}{\underset{b}{\rightrightarrows}} X \xrightarrow{i} Y$$

such that $ai = bi$ then already $a = b$.

Show that in the category of sets, the monomorphisms are precisely the injective maps.

3. An arrow $p : A \to B$ in a category $\boldsymbol{C}$ is called an *epimorphism* if it is cancellable from the left: this means that given a diagram

$$A \xrightarrow{p} B \overset{f}{\underset{g}{\rightrightarrows}} Z$$

such that $pf = pg$ then already $f = g$.
Show that in the category of sets, the epimorphisms are precisely the surjective maps.

The notions of monomorphism and epimorphism are more general and more widely applicable than injections and surjections, which only make sense in categories whose objects have elements.

A.2 Functors

The feature of categories as opposed to sets, is the emphasis on arrows: e.g. the study of groups is really the study of group homomorphisms, etc. By the same principle, in order to study categories we should study their 'maps'.

A.2.1 Functors. A functor F between two categories $\boldsymbol{C}$ and $\boldsymbol{D}$ consists of

- a map F from $\boldsymbol{C}_0$ to $\boldsymbol{D}_0$,
- for each pair of objects X, Y in $\boldsymbol{D}$, a map $F_{X,Y} : \boldsymbol{C}(X, Y) \to \boldsymbol{D}(XF, YF)$, preserving composition law and identity arrows.

To preserve the composition means that given arrows $X \xrightarrow{f} Y \xrightarrow{g} Z$ in $\boldsymbol{C}$ we have $(fF)(gF) = (fg)F$ in $\boldsymbol{D}$. And for every object X in $\boldsymbol{C}$, the map $F_{X,X}$ sends Id_X to the identity arrow of XF in $\boldsymbol{D}$.

A.2.2 Examples of functors. *Subcategories* provide examples of functors: for example ***FinSet*** $\to$ ***Set*** is the functor that to a finite set assigns the same set, ignoring that it happens to be finite.

Forgetful functors discard structure: for example there is a forgetful functor ***Vect***$_{\Bbbk}$ $\to$ ***Set*** which to each vector space V associates the underlying set (the set of vectors in V), and to each linear map associates the underlying set map. It is clear that this assignment respects composition and identity arrows.

Similarly there are forgetful functors ***Top*** $\to$ ***Set***, ***Grp*** $\to$ ***Set***, ***FinOrd*** $\to$ ***FinSet***, etc. In this last example, note that although every finite ordered set is a finite set, we cannot say that ***FinOrd*** is a subcategory in ***FinSet***, because each particular set admits many orderings.

As an example of a functor 'in the other direction', consider the functor $\boldsymbol{Set} \to \boldsymbol{Vect}_{\mathbb{k}}$ which to each set associates the vector space spanned by the elements (i.e. the vector space of all formal linear combinations of the elements in S). On arrows: given a set map $S \to T$, the associated linear map is the one given by extending by linearity.

A.2.3 The opposite category, and contravariant functors. Given a category $\boldsymbol{C}$, the opposite category $\boldsymbol{C}^{\text{op}}$ is given by taking the same objects but reversing the direction of all arrows, that is, $\boldsymbol{C}^{\text{op}}(X, Y) = \boldsymbol{C}(Y, X)$.

A functor $\boldsymbol{C}^{\text{op}} \to \boldsymbol{D}$ is called a contravariant functor. Compared to a usual (covariant) functor $\boldsymbol{C} \to \boldsymbol{D}$ it reverses the direction of all arrows. For example, there is a functor $\boldsymbol{Top}^{\text{op}} \to \boldsymbol{Ring}$ which to a topological space X associates the ring $C(X)$ of all continuous functions $X \to \mathbb{R}$. The contravariance comes about because given a continuous map $X \to Y$, there is a ring homomorphism $C(Y) \to C(X)$ given by sending $Y \to \mathbb{R}$ to the composite $X \to Y \to \mathbb{R}$.

For any category there is always the *identity functor* $\text{Id}_{\boldsymbol{C}} : \boldsymbol{C} \to \boldsymbol{C}$ which is the identity map on objects and the identity map on arrows.

Also *we can compose two functors* $F : \boldsymbol{C} \to \boldsymbol{D}$ and $G : \boldsymbol{D} \to \boldsymbol{E}$, to get a functor $FG : \boldsymbol{C} \to \boldsymbol{E}$. This is defined by composing F_0 with G_0, and for each $X, Y \in \boldsymbol{C}_0$ compose $F_{X,Y}$ with $G_{XF,YF}$. It is easy to check that this composition law satisfies associativity, and also that the identity functors are neutral for this composition law, from both sides.

The notions of categories and functors fit together to give

A.2.4 The category of all categories. The category $\boldsymbol{Cat}$ is the category whose objects are the categories, and whose arrows are the functors.

(Here we ignore some set-theoretical subtleties: just as there is no such thing as the set of all sets (which would lead to Russell's paradox), there is really no such thing as the category of all categories... There are several ways to deal with this problem, and there is no point in bothering about this in our context.)

A functor is called an *isomorphism* of categories if there is a functor in the other direction such that the two composites are both equal to the identity functors.

Next, the gadget which allows for comparison of functors is

A.2.5 Natural transformations. Given two categories $\boldsymbol{C}$ and $\boldsymbol{D}$, and two functors

$$\boldsymbol{C} \underset{F}{\overset{G}{\rightrightarrows}} \boldsymbol{D}$$

a *natural transformation* $u : F \Rightarrow G$ is the data of

- For each object $X \in \boldsymbol{C}_0$ an arrow $u_X : XF \to XG$ in $\boldsymbol{D}$ (called the components of u).

These maps must be *natural*, i.e. compatible with all arrows in $\boldsymbol{C}$: for every arrow $\alpha : X \to Y$ in $\boldsymbol{C}$, this diagram must commute:

$$\begin{array}{ccc} XF & \xrightarrow{u_X} & XG \\ {\scriptstyle \alpha F}\downarrow & & \downarrow{\scriptstyle \alpha G} \\ YF & \xrightarrow[u_Y]{} & YG \end{array}$$

There is always the notion of *identity natural transformation* between a given functor and itself. It is simply the one given by taking all the components to be the identity arrows. Also *one can compose natural transformations*: this is just a matter of composing its components. It is not difficult to see that this composition law is associative, and that the identity natural transformations are neutral for this composition.

A *natural isomorphism of functors* is a natural transformation which admits a natural transformation in the other direction such that the two compositions are equal to the identity natural transformations.

Now, functors and natural transformations together lead to the notion of

A.2.6 Functor categories. For two fixed categories $\boldsymbol{C}$ and $\boldsymbol{D}$ we can consider the set of all functors $\boldsymbol{C} \to \boldsymbol{D}$. These constitute the object set of a category denoted $\boldsymbol{Cat}(\boldsymbol{C}, \boldsymbol{D})$. The arrows are all natural transformations between such functors.

A.2.7 Example. Consider the category $\boldsymbol{Cat}(\mathbf{2}, \boldsymbol{Set})$, where $\mathbf{2}$ denotes the category $\{0 \to 1\}$ mentioned in A.1.4. A functor $\mathbf{2} \to \boldsymbol{Set}$ is given by specifying two sets S_0 and S_1 (the images of the objects 0 and 1 in $\mathbf{2}$), and an arrow $S_0 \to S_1$ (the image of the unique arrow in $\mathbf{2}$) (and then by definition the identity arrows are required to be sent to identity arrows, so there is no choice to be made for them). Let there now be given two functors, i.e. two objects in $\boldsymbol{Cat}(\mathbf{2}, \boldsymbol{Set})$: one given by $S_0 \to S_1$, and another given by $T_0 \to T_1$. To give a natural transformation $u : S \Rightarrow T$ between these two functors amounts to giving maps $u_0 : S_0 \to T_0$ and $u_1 : S_1 \to T_1$, and the naturality requirement is that this square commutes:

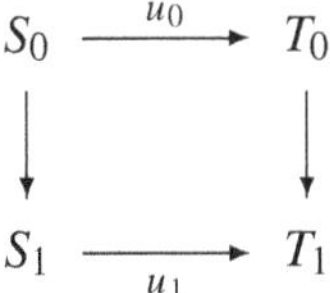

Altogether we have described the functor category $\boldsymbol{Cat}(\mathbf{2}, \boldsymbol{Set})$ as the category whose objects are set maps, and whose arrows are commutative squares.

Many more examples of natural transformations and functor categories are given in the main text.

A.2.8 Faithfulness, fullness, equivalences. A functor $F : \boldsymbol{C} \to \boldsymbol{D}$ is called *faithful* if for each pair of objects $X, Y \in \boldsymbol{C}_0$ the map $F_{X,Y} : \boldsymbol{C}(X, Y) \to \boldsymbol{D}(XF, YF)$ is injective. F is called *full* if the maps $F_{X,Y} : \boldsymbol{C}(X, Y) \to \boldsymbol{D}(XF, YF)$ are all surjective. If $\boldsymbol{C}$ is a subcategory of $\boldsymbol{D}$ then the inclusion functor is always faithful. If it is full, $\boldsymbol{C}$ is called a *full subcategory*.

A functor is called *essentially surjective* if every object in $\boldsymbol{D}$ is isomorphic to an image under F of an object of $\boldsymbol{C}$.

A functor is called an *equivalence* if it is faithful, full, and essentially surjective. An equivalent characterisation is that there exists a functor in the other direction which is almost an inverse. The precise formulation of this last condition is that the composite (of functors) $FG : \boldsymbol{C} \to \boldsymbol{C}$ is naturally isomorphic to the identity functor $\mathrm{Id}_{\boldsymbol{C}}$, and that $GF : \boldsymbol{D} \to \boldsymbol{D}$ is naturally isomorphic to the identity functor $\mathrm{Id}_{\boldsymbol{D}}$.

Exercises

1. In Example A.1.3 we assigned a one-object category $\boldsymbol{G}$ to every group G. Show how to assign a functor $\boldsymbol{G} \to \boldsymbol{H}$ to every group homomorphism $G \to H$. Show that all this together defines a functor from $\boldsymbol{Grp}$ to $\boldsymbol{Cat}$, and that this functor is faithful and full.
2. Show that the forgetful functors listed in A.2.2 are all faithful but not full.
3. Let $F : \boldsymbol{C} \to \boldsymbol{D}$ be a faithful functor, and let a be an arrow in $\boldsymbol{C}$. Show that if aF is a monomorphism (respectively an epimorphism) then already a is a monomorphism (respectively an epimorphism).
4. Show that a natural transformation $u : F \Rightarrow G$ between functors as above is a natural isomorphism if and only if all of its components are invertible arrows in $\boldsymbol{D}$.

A.3 Universal objects

A.3.1 Initial and terminal objects. An object I in a category $\boldsymbol{C}$ is called an *initial object* if for every object $X \in \boldsymbol{C}_0$ there is precisely one arrow $I \to X$. In particular, if I is an initial object, then there is only one arrow $I \to I$ (and this arrow is then of course the identity arrow).

If I and J are both initial objects, then there is a unique arrow $i : I \to J$ and a unique arrow $j : J \to I$. The composite ij is an arrow from I to I, so it must be the identity, and for the same reason ji must be the identity of J. So in conclusion, between any two initial objects there is a unique isomorphism.

Dually, an object T is called a *terminal object* if for every object X there is precisely one arrow $X \to T$. Arguing the same way, we see that between any two terminal objects there is a unique isomorphism. (You can say that the terminal objects of $\boldsymbol{C}$ are just the initial objects in $\boldsymbol{C}^{\mathrm{op}}$.)

A.3.2 Examples. In the categories ***Set***, ***Vect***$_{\Bbbk}$, ***Top*** and ***Cat***, any object with just one element is a terminal object. In ***Set***, ***Top***, and ***Cat***, the empty object is initial. In ***Vect***$_{\Bbbk}$, the trivial vector space $\{0\}$ is initial (as well as terminal). In the category $\boldsymbol{G}$ associated to a nontrivial group, there is only one object, but it is neither initial nor terminal, because there is more than one arrow from this object to itself.

A.3.3 Products. Let $\boldsymbol{C}$ be a category and let X and Y be two objects. A (categorical) product of X and Y is an object P equipped with two arrows $X \longleftarrow P \longrightarrow Y$ called projections, with the following universal property: for every other object Q equipped with arrows $X \longleftarrow Q \longrightarrow Y$ there is a unique arrow $Q \to P$ which makes this diagram commute:

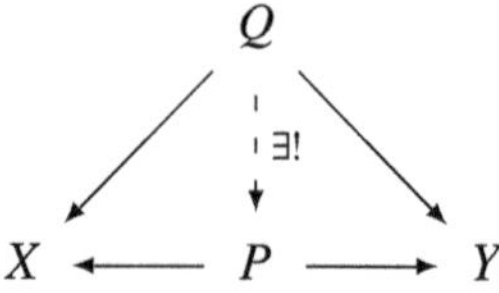

You can formulate this by defining a category whose objects are diagrams $\underline{A}$ of type $X \longleftarrow A \longrightarrow Y$, and whose arrows $\underline{A} \to \underline{B}$ are diagrams like this:

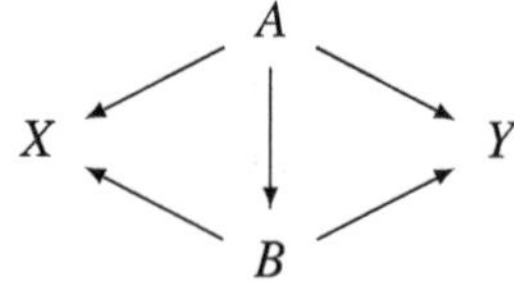

Now the product can be characterised as the terminal object in this category. (This category is nothing but the functor category $\boldsymbol{Cat}(\boldsymbol{P}, \boldsymbol{C})$ where $\boldsymbol{P}$ denotes the category $\bullet \leftarrow \bullet \rightarrow \bullet$, cf. Definition A.2.6, in analogy with Example A.2.7.)

If a product of X and Y exists then it is unique up to unique isomorphism. This means that if two objects P and P' are both products of X and Y, then there is a unique isomorphism between them compatible with the projections. (This follows easily from the universal property of P and P'.)

In many categories products exist for any two objects. For example, in each of the categories ***Set***, ***Vect***$_{\Bbbk}$, ***Top***, ***Cat***, the usual cartesian product is a categorical product. (Note that in ***nCob*** there is no categorical product.)

A.3.4 Coproducts. Dually, a coproduct of X and Y is an object C with two arrows $X \longrightarrow C \longleftarrow Y$ which is universal among all such diagrams. It can also be described as the objects which are products in the opposite category, or as the initial objects in the category of such diagrams.

In each of the categories ***Set***, ***Top***, ***Cat***, the operation of *disjoint union* is a coproduct – this operation is described in great detail in 1.3.24. In the category ***Vect***$_{\Bbbk}$, the direct sum is a coproduct (as well as a product).

A.3.5 n-ary products. More generally, given n objects $X_1, \ldots, X_n$, their product is an object P with maps to each X_i, universal among such diagrams. If the (binary) products exist, denoted $X \times Y$, then it is not difficult to show that $(X \times Y) \times Z$ and $X \times (Y \times Z)$ are both categorical triple products (hence canonically isomorphic).

You can also consider the empty product! It is an object T equipped with arrows to each of the zero objects, such that for every other object equipped with such maps (none at all) there is a unique arrow to T (compatible with those zero structure arrows). In other words, it is precisely a terminal object!

A.3.6 Limits and fibre products. One can consider more complicated diagrams than just those which consist of an object and a collection of arrows to several other objects. This leads to the more general notion of limits. We will need a particular case: the fibre product. Suppose we are given two objects X and Y, each with a specified arrow to Z, i.e. a diagram

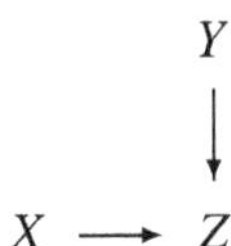

A *fibre product* F is an object with arrows to each of those objects, i.e. a commutative diagram

$$\begin{array}{ccc} F & \longrightarrow & Y \\ \downarrow & & \downarrow \\ X & \longrightarrow & Z \end{array}$$

such that for every other such diagram

$$\begin{array}{ccc} G & \longrightarrow & Y \\ \downarrow & & \downarrow \\ X & \longrightarrow & Z \end{array}$$

there is a unique arrow $G \to F$ which makes this diagram commute:

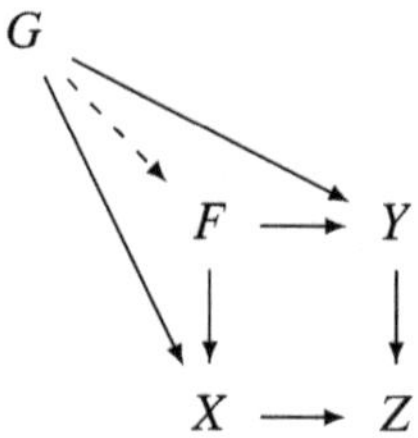

A.3.7 Pushouts. Finally there is of course the dual notion. Given two objects X and Y, each with a specified arrow from Z, i.e. a diagram

$$\begin{array}{ccc} & & Y \\ & & \uparrow \\ X & \longleftarrow & Z \end{array}$$

A *pushout* P is an object with arrows from each of those objects, i.e. a commutative diagram

$$\begin{array}{ccc} P & \longleftarrow & Y \\ \uparrow & & \uparrow \\ X & \longleftarrow & Z \end{array}$$

which is universal among such diagrams, i.e. for any other such diagram

$$\begin{array}{ccc} Q & \longleftarrow & Y \\ \uparrow & & \uparrow \\ X & \longleftarrow & Z \end{array}$$

there is a unique arrow $Q \leftarrow P$ which makes this diagram commute:

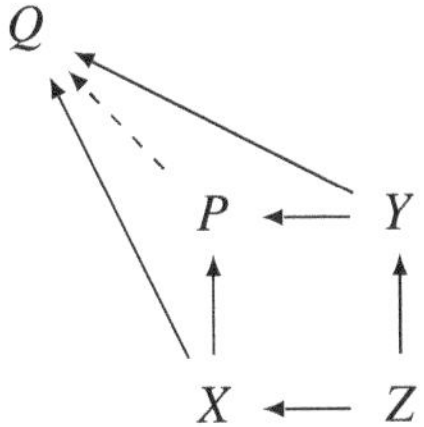

Exercises

1. Show that the category $\mathbf{2} = \{0 \to 1\}$ has an initial and a terminal object, and that it has products and coproducts.
2. Let $\boldsymbol{G}$ denote the category associated to a nontrivial group. Show that products do not exist in $\boldsymbol{G}$. (There is only one object in this category, say x, so the only case to check is whether the product of x with itself exists. The only candidate for being a product is the object x. But there are many candidates for being the projection maps, and the exercise consists in showing that none of them gives the universal property.)
3. Given set maps $X \xrightarrow{f} Z \xleftarrow{g} Y$, show that the set
$$X \times_Z Y := \{(x, y) \in X \times Y \mid xf = yg\}$$
(with its two projections induced from $X \times Y$) is a fibre product in ***Set***.
4. Let ***Incl*** denote the category whose objects are sets and whose arrows are inclusions of sets. Given a diagram $A \supset Z \subset B$, show that the plain union $A \cup B$ (with the natural inclusions of A and B) is a pushout in ***Incl***.

References

[1] LOWELL ABRAMS. Two-dimensional topological quantum field theories and Frobenius algebras. *J. Knot Theory Ramifications* **5** (1996), 569–587. Available at http://home.gwu.edu/~labrams/docs/tqft.ps.

[2] LOWELL ABRAMS. Modules, comodules, and cotensor products over Frobenius algebras. *J. Algebra* **219** (1999), 201–213. Available at http://home.gwu.edu/~labrams/docs/cotensor.ps.

[3] LOWELL ABRAMS. The quantum Euler class and the quantum cohomology of the Grassmannians. *Isr. J. Math.* **117** (2000), 335–352. Available at http://home.gwu.edu/~labrams/docs/qeuler.ps.

[4] HANS CHRISTIAN ANDERSEN. *Fyrtøjet*. København, 1835. (English translation, *The Tinder-Box*, available on the internet.)

[5] MICHAEL ATIYAH. Topological quantum field theories. *Inst. Hautes Études Sci. Publ. Math.* **68** (1989), 175–186.

[6] MICHAEL ATIYAH. *The Geometry and Physics of Knots*. Cambridge University Press, Cambridge, 1990.

[7] MICHAEL ATIYAH. An introduction to topological quantum field theories. *Turk. J. Math.* **21** (1997), 1–7.

[8] JOHN BAEZ's web site, http://math.ucr.edu/home/baez/.

[9] JOHN BAEZ and JAMES DOLAN. Higher-dimensional algebra and topological quantum field theory. *J. Math. Phys.* **36** (1995), 6073–6105 (q-alg/9503002).

[10] JOHN BAEZ and JAMES DOLAN. From finite sets to Feynman diagrams. In B. Engquist and W. Schmid, editors, *Mathematics unlimited – 2001 and beyond*, pp. 29–50. Springer-Verlag, Berlin, 2001 (math.QA/0004133).

[11] JOHN W. BARRETT. Quantum gravity as topological quantum field theory. *J. Math. Phys.* **36** (1995), 6161–6179 (gr-qc/9506070).

[12] RAOUL BOTT and LORING W. TU. *Differential Forms in Algebraic Topology*. No. 82 in Graduate Texts in Mathematics. Springer-Verlag, New York, 1982.

[13] HAROLD S. M. COXETER and WILLIAM O. MOSER. *Generators and Relations for Discrete Groups*. No. 14 in Ergebnisse der Mathematik und ihrer Grenzgebiete. Springer-Verlag, 1957.

[14] LOUIS CRANE and DAVID YETTER. On algebraic structures implicit in topological quantum field theories. *J. Knot Theory Ramifications* **8** (1999), 125–163 (hep-th/9412025).

[15] CHARLES W. CURTIS and IRVING REINER. *Representation Theory of Finite Groups and Associative Algebras*. Interscience Publishers, New York, 1962.

[16] ROBBERT DIJKGRAAF. A geometric approach to two dimensional conformal field theory. Ph.D. Thesis, University of Utrecht, 1989.

[17] ROBBERT DIJKGRAAF. Fields, strings and duality. In *Symétries Quantiques (Les Houches, 1995)*, pp. 3–147. North-Holland, Amsterdam, 1998 (hep-th/9703136).

[18] ROBBERT DIJKGRAAF, ERIK VERLINDE, and HERMAN VERLINDE. Notes on topological string theory and 2D quantum gravity. In M. Green *et al.*, editors, *Proceedings of the Trieste Spring School, 1990: String Theory and Quantum Gravity*, pp. 91–156. World Scientific, Singapore, 1991.

[19] BORIS DUBROVIN. Geometry of 2D topological field theories. In *Integrable Systems and Quantum Groups*, no. 1620 in Lecture Notes in Mathematics, pp. 120–348. Springer-Verlag, New York, 1996 (hep-th/9407018).

[20] BERGFINNUR DURHUUS and THÓRDUR JÓNSSON. Classification and construction of unitary topological field theories in two dimensions. *J. Math. Phys.* **35** (1994), 5306–5313 (hep-th/9308043).

[21] DAVID EISENBUD. *Commutative Algebra with a View toward Algebraic Geometry*. No. 150 in Graduate Texts in Mathematics. Springer-Verlag, New York, 1995.

[22] GEORG FROBENIUS. Theorie der hyperkomplexen Grössen. *Sitzungsber. K. Preuss. Akad. Wiss.* **24** (1903), 504–537; 634–645.

[23] WILLIAM FULTON. *Algebraic Topology: A First Course*. No. 153 in Graduate Texts in Mathematics. Springer-Verlag, New York, 1995.

[24] PAUL G. GOERSS and JOHN F. JARDINE. *Simplicial Homotopy Theory*. Vol. 174 of Progress in Mathematics. Birkhäuser, Basel, 1999.

[25] PHILLIP GRIFFITHS and JOE HARRIS. *Principles of Algebraic Geometry*. Wiley, 1978.

[26] M. Hazewinkel, editor. *Encyclopaedia of Mathematics*. Vol. 1–6. A–Zyg, index. Kluwer Academic Publishers, Dordrecht, 1995. Translated from the Russian, Reprint of the 1988–1994 English translation.

[27] MORRIS HIRSCH. *Differential Topology*. No. 33 in Graduate Texts in Mathematics. Springer-Verlag, New York, 1976.

[28] ANDRÉ JOYAL and ROSS STREET. Braided tensor categories. *Adv. Math.* **102** (1993), 20–78.

[29] CHRISTIAN KASSEL. *Quantum Groups*. No. 155 in Graduate Texts in Mathematics. Springer-Verlag, New York, 1995.

[30] SERGE LANG. *Algebra*. Addison-Wesley, Reading, MA, 1971.

[31] H. BLAINE LAWSON JR. *The Theory of Gauge Fields in Four Dimensions*. CBMS Regional Conference Series in Mathematics, **58**, Published for the Conference Board of the Mathematical Sciences, Washington, DC, 1985.

[32] F. WILLIAM LAWVERE. Ordinal sums and equational doctrines. In B. Eckmann, editor, *Seminar on Triples and Categorical Homology Theory, ETH 1966/67*. No. 80 in Lecture Notes in Mathematics, pp. 141–155. Springer-Verlag, New York, 1967.

[33] F. WILLIAM LAWVERE. Metric spaces, generalized logic, and closed categories. *Rend. Semin. Mat. Fis. Milano* **43** (1973), 135–166. Reprinted in *Repr.*

Theory Appl. Categ. **1** (2002), 36pp. (electronic), with an author commentary: Enriched categories in the logic of geometry and analysis. Available at http://www.tac.mta.ca/tac/reprints/articles/1/tr1abs.html.

[34] SAUNDERS MAC LANE. *Categories for the Working Mathematician*, second edition. No. 5 in Graduate Texts in Mathematics. Springer-Verlag, New York, 1998.

[35] JOHN MILNOR. On manifolds homeomorphic to the 7-sphere. *Ann. Math.* **64** (1956), 399–405.

[36] JOHN MILNOR. *Lectures on the h-Cobordism Theorem.* Princeton University Press, Princeton, NJ, 1965.

[37] EDWIN E. MOISE. *Geometric Topology in Dimensions 2 and 3.* No. 47 in Graduate Texts in Mathematics. Springer-Verlag, New York, 1977.

[38] ELIAKIM HASTINGS MOORE. Concerning the abstract groups of order $k!$ and $\frac{1}{2}k!$ holohedrically isomorphic with the symmetric and the alternating substitution-groups on k letters. *London Soc. Math. Proc.* **28** (1897), 357–366.

[39] MICHAEL MÜGER. From subfactors to categories and topology I. Frobenius algebras in and Morita equivalence of tensor categories. *J. Pure Appl. Alg.* **180** (2003), 81–157 (math.CT/0111204).

[40] JAMES MUNKRES. *Elementary Differential Topology.* No. 54 in Annals of Mathematics Study. Princeton University Press, Princeton, NJ, 1963.

[41] TADASI NAKAYAMA. On Frobeniusean algebras I. *Ann. Math.* **40** (1939), 611–633.

[42] CECIL NESBITT. On the regular representations of algebras. *Ann. Math.* **39** (1938), 634–658.

[43] FRANK QUINN. Lectures on axiomatic topological quantum field theory. In *Geometry and Quantum Field Theory (Park City, UT, 1991)*, pp. 323–453. Amer. Math. Soc., Providence, RI, 1995.

[44] STEPHEN SAWIN. Direct sum decompositions and indecomposable TQFTs. *J. Math. Phys.* **36** (1995), 6673–6680 (q-alg/9505026).

[45] GRAEME SEGAL. Two-dimensional conformal field theories and modular functors. In *IXth International Congress on Mathematical Physics (Swansea, 1988)*, pp. 22–37. Hilger, Bristol, 1989.

[46] MOSS E. SWEEDLER. *Hopf algebras.* W. A. Benjamin, New York, 1969.

[47] RENÉ THOM. Quelques propriétés globales des variétés différentiables. *Comment. Math. Helv.* **28** (1954), 17–86.

[48] ANDREW H. WALLACE. *Differential Topology: First Steps.* W. A. Benjamin, New York, 1968.

[49] JOSEPH H. M. WEDDERBURN. On hypercomplex numbers. *Proc. London Math. Soc.* **6** (1908), 77–118.

[50] EDWARD WITTEN. Topological quantum field theory. *Commun. Math. Phys.* **117** (1988), 353–386.

Index

For EU product safety concerns, contact us at Calle de José Abascal, 56–1°, 28003 Madrid, Spain or eugpsr@cambridge.org.

www.ingramcontent.com/pod-product-compliance
Ingram Content Group UK Ltd.
Pitfield, Milton Keynes, MK11 3LW, UK
UKHW042209080726
473066UK00007B/324

* 9 7 8 0 5 2 1 5 4 0 3 1 5 *